# HANDBOOK OF SEED SCIENCE

# HANDBOOK OF SEED SCIENCE

R. V. RAMANAN
REKHA GOGOI

KNOWLEDGE
BAKERS

KNOWLEDGE BAKERS

ISBN: 978-93-49964-85-3



*Published by* : **Knowledge Bakers**
Pune, India
*info@knowledgebakers.com*

# PREFACE

Handbook of Seed Science presents a comprehensive exploration of the biological, physiological, ecological, and technological principles that govern seed development, viability, storage, and utilization. Serving as a foundational and applied reference, this volume brings together key concepts and contemporary research that underpin seed science, a discipline central to agriculture, biodiversity conservation, plant breeding, and food security.

Seeds represent the beginning of the plant life cycle and play a critical role in the propagation of plant species. Their quality, performance, and resilience directly influence crop productivity and ecological restoration efforts. The scientific study of seeds encompasses multiple dimensions—from the molecular and genetic regulation of embryogenesis to the practical considerations of seed testing, certification, and long-term storage.

This handbook is organized to provide a systematic and detailed treatment of the subject. Initial chapters address the structure and development of seeds, covering processes such as fertilization, embryogenesis, and endosperm formation. These sections also examine seed dormancy and germination mechanisms, with attention to the hormonal, genetic, and environmental cues that regulate these vital stages.

Following this, the book delves into seed physiology and biochemistry, discussing topics such as desiccation tolerance, metabolic activity during germination, and seed aging. Advanced techniques for evaluating seed vigor, viability, and health are also discussed, alongside internationally accepted testing standards and protocols.

A significant portion of the volume is devoted to seed technology, including harvesting methods, cleaning, drying, and treatment techniques aimed at enhancing seed quality and protecting against pathogens. The principles of seed storage are covered extensively, with a focus on maintaining genetic integrity and viability over time in both ex situ conservation and commercial contexts.

Chapters on seed certification and legislation provide an overview of regulatory frameworks governing seed trade, quality control, and intellectual property rights, while highlighting global initiatives and harmonization efforts. The importance of seed banks, gene banks, and biotechnological innovations in conservation and breeding programs is also addressed.

Recent advancements in seed science—such as molecular seed diagnostics, priming techniques, and nanotechnology applications—are presented to illustrate the evolving tools and methodologies that enhance precision and sustainability in seed-related practices.

Throughout the handbook, real-world applications are emphasized, linking scientific theory with agricultural innovation, ecological restoration, and climate resilience. Interdisciplinary perspectives ensure that readers from plant science, agronomy, environmental science, and biotechnology will find relevance and utility in the content.

This volume aims to support researchers, students, practitioners, and policymakers engaged in seed science and related fields. By synthesizing foundational knowledge with emerging trends, Handbook of Seed Science aspires to contribute meaningfully to global efforts in securing crop diversity, ensuring food security, and advancing sustainable agriculture.

We extend our sincere gratitude to the scholars, institutions, and professionals whose expertise and dedication have enriched this volume. It is our hope that this handbook serves as both a scholarly resource and a practical guide for advancing the science and stewardship of seeds.

**R. V. Ramanan**

**Rekha Gogoi**

# CONTENTS

ORNAMENTAL CROPS

# Chapter 1

# TECHNIQUES OF SEED PRODUCTION AND HANDLING

## INTRODUCTION

More plants are propagated for food, fiber, and ornamentals from seeds than any other method of propagation. Seed propagation is the cornerstone for producing agronomic, vegetable, forestry, and many ornamental plants. The production of high-quality seeds is of prime importance to propagators. In the production of any crop, the cost of the seed is usually minor compared with other production costs, yet, no single factor is as important in determining the success of the operation. Most crop plant seeds are produced by companies that specialize in both plant breeding and seed production. Growers expect these companies to introduce improved cultivars, as well as to produce high-quality seeds that have good germination characteristics and are true-to-type.

To produce high-quality seeds, companies must not only pay close attention to the environment where seeds are produced, but must also have the means to test the quality of those seeds. This chapter discusses various aspects of seed production, testing, and storage. The steps taken to produce, clean, and store seeds for commercial crop production are summarized for a variety of crops in Table -1.

## Table-1

**Steps for Producing, Cleaning, and Storage of Seeds for Commercial Crop Production**

| Crop | Production practices | Seed conditioning | Seed storage | Seed treatments |
|---|---|---|---|---|
| Sweet corn | Hybrid seed production from two inbred parents by wind pollination. Female parent requires detasseling before pollen is shed and is interplanted with rows of the male pollen parent. | Corn cobs are harvested when the seeds are between 35 and 45% moisture to avoid mechanical injury during harvest. Cobs are force-air–dried to about 12 to 13% moisture where the seeds are mechanically removed from the cob. Final moisture is removed in a drying oven (35 to 40°C). | Stored at 10% moisture at 10°C. | Usually treated with fungicide and/or insecticide. Often applied in a polymer film coating. |
| Tomato | Hybrid seed from inbred parents by hand pollination. Seed parent may be male-sterile, or hand emasculation of anthers is required. | Fruit pulp is separated from the seeds by juice extracting equipment. Extracts can be fermented for 2 to 3 days until the seeds separate from fruit gel and sink. Treatment with HCl acid (5%) is also used to extract seeds after several hours. Excessive fermentation or chemical treatment reduces seed quality. Seed drying should not exceed 43°C. | Stored at 6% moisture at 5 to 10°C. | Can be treated with a fungicide or, in some cases, primed. |
| Onion | Hybrid onion seed is produced by insect pollination between inbred parents. The female seed parent is male-sterile. Plants flower (bolt) after the second season. It is common to plant seed at close spacing the first year to produce small bulbs that are replanted at the appropriate spacing for seed production the second spring. | Seed maturity can vary because flowering umbels are not all initiated at the same time on the plant. Harvest the entire umbel when the first individual fruits begin to crack and show black seeds. Umbels are naturally air-dried for 2 to 3 weeks on open benches. These are threshed and seeds are separated by screens, air, and gravity separation. | Seeds of onion are short-lived in storage. Stored at 6% moisture at 5°C. | No special seed treatments. |
| Impatiens | Hybrid seeds are produced in the greenhouse by hand pollination between inbred parents. Seed parent is pollinated as soon as the stigma is receptive to prevent self-pollination. | Fruit of impatiens explodes when ripe, expelling seeds. Therefore, fruits are harvested prior to expulsion and placed on frames for several days until seeds are shed. Seeds are then air-dried or dried under gentle heat. | Stored between 3 and 5% moisture at 5°C. | Impatiens are a high-value seed crop. Seeds may be primed, pelleted, or pregerminated. |
| Pawpaw (*Asimina*) | Pawpaw understocks are produced from seeds. Hand pollination between trees with different genetic backgrounds will increase fruit and seed set. | In most cases, seeds are a by-product of fruit processing. Pulp can be removed by fermentation and washing. | Pawpaw seeds are recalcitrant and cannot withstand seed moisture below 35%. Seeds can be stored moist at 5°C for 2 years. | Stratification (moist, cold storage) for 8 to 10 weeks to relieve dormancy. |
| Pine | Seed orchards are established with elite trees with superior growth characteristics. Seed production takes 18 months and trees take between 2 and 10 years to bear a crop. | For some species, seed is collected on nets under trees after the cones naturally shed seeds. For most, cones are harvested and placed on wire benches where the cones air dry and shed seeds in 2 to 8 weeks. Some cones require oven drying at about 50°C to open cones. Seeds are collected and mechanically dewinged, followed by flotation or gravity separation to get viable seeds. | Stored at 6% moisture and 0 to 5°C. | Stratification (moist, cold storage) for 2 to 12 weeks to relieve dormancy. |

## Sources for Seeds

### Commercial Seed Production

Commercial seed production is a specialized intensive industry with its own technology geared to the requirements of individual species (Fig. 1,). This section on sources for seeds will be separated into herbaceous and woody plant seeds.

Historically, seeds for next season's crop were collected as a by-product of production. Although some seeds may still be produced in this manner (e.g., some Third World production), modern seed production has become a very specialized industry. A scheme for producing quality seed is included in Figure –2.

Some agricultural seeds—such as corn, wheat, small grains, and grasses—are produced in the area where the crops are grown. The advantages for producing seeds in their production area include reduced transportation and handling costs as well as reduced potential for genetic shift. These are important considerations for agronomic crops where large amounts of seeds are required to produce a crop. However, crop production areas may not provide the best conditions for producing high-quality, disease-free seeds.

*Figure –1 A majority of important agronomic, horticultural, and forestry crops are propagated by seeds that come in a large diversity of seed size and shape, resulting in diverse requirements for seed production, extraction, and conditioning.*

Therefore, large amounts of high-value seeds such as forage, vegetable, and flowers are produced in specialized growing areas. The major considerations for selecting areas to produce seeds are environmental conditions and cost of production. Large quantities of grass, vegetable, and flower seeds are produced in areas characterized by low summer rainfall, low humidity, and limited rainfall during the seed harvest season. These conditions provide good seed yields and reduce disease problems, especially during harvest when seeds must dry before being handled. There are also crops that require special environmental conditions to flower and set seeds. These include the biennial vegetable and flower crops that require **vernalization** (a period of cold temperature) to flower. Examples are onion and carrot seed production.

One-year-old biennial plants used for seed production have been called **stecklings.**

Plants may be chilled by overwintering in the field, or in some cases, stecklings are brought into a cooler (5°C, 40°F) to satisfy vernalization requirements and shorten the seed-production cycle.

Major production areas for high-value seed production in the United States that meet these important environmental conditions include grass and forage seed production in the Pacific Northwest and vegetable and flower seed production in the Pacific Northwest down to the central, coastal valleys of California (Fig.–3). Increasingly, seed production has become an international industry. For example, the United States, Netherlands, and Japan provide over half of the world's flower seeds. Hybrid seed production that requires hand pollination has moved to areas of the world with reduced labor costs. These include Central and South America, Southeast Asia, India, and Africa. The advantages to producing seeds in the Southern Hemisphere include a reduced cost of production, and seed production in the season prior to planting in northern crop production areas, which reduces storage time and cost.

Regardless of the country where seeds are produced, there are several important considerations that must be satisfied when selecting specific sites for **seed production** :

1. Appropriate **soil type** and **fertility** for good seed yields.
2. A detailed **cropping history** to avoid disease or herbicide carryover.
3. Adequate **soil moisture** or availability of supplemental irrigation.
4. A **dry environment** during seed harvest.
5. Ability to isolate **open or cross-pollinated crops.**

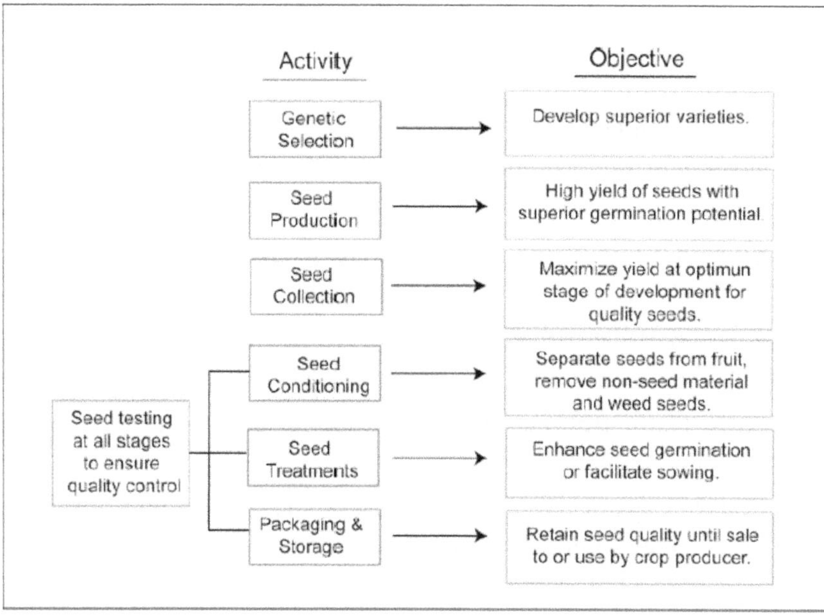


*Figure -2 Procedures for producing and handling a commercial seed lot.*

*Figure -3 Seed production fields (a) Mallow produced as wildflower seed in Oregon. (b) Wildflowers (coneflower in forefront and grasses behind) production in Wisconsin. California production of (c) cucumber and (d) sunflower with bee hives for pollination.*

For example, self-pollinated tomato plants require only 50 feet of separation between varieties, while some insect- or wind-pollinated crops require up to a mile of separation between varieties to avoid unwanted cross-pollination.

Additional requirements for high-quality seed production are the selection of planting density, pest control, and availability of insect pollinators. In many cases, conditions for seed production and crop production are very similar.

**Woody Plant Seed** A number of commercial and professional seed-collecting firms exist that collect and sell seeds of certain timber, ornamental, or fruit species.

Lists of such producers are available. Such seeds should be properly labeled as to their origin or provenance. Some tree seeds can be obtained as certified seeds.

Seed Exchanges. Many arboreta and plant societies have seed exchanges or will provide small amounts of specialty seed.

Seed Collecting. Propagators at individual nurseries may collect tree and shrub seeds. These may be collected from specific seed-collection zones or from seed-production areas.

Seeds may be collected from standing trees, trees felled for logging, or from squirrel caches. They might be collected from parks, roadways, streets, or wood lots. Seed collecting has the advantage of being under the control of the propagator, but requires intimate knowledge of each species and the proper method of handling.

Seed Orchards. Seed orchards or plantations are used to maintain seed source trees of particularly valuable species. They are extensively used by nurseries in the production of rootstock seeds of certain species or cultivars and for forest tree improvement. The major advantage to a seed orchard is that it is a consistent source of seeds from a known (often genetically superior) parentage. They also allow the seed producer to maximize seed harvest by reducing loss due to environmental conditions or animals.

Fruit-Processing Industries. Historically, many of the fruit tree rootstock seeds were obtained as by-products of fruit-processing industries such as canneries, cider presses, and dry yards. Examples include peach and apricot in California, as well as pears in the Pacific Northwest. The procedure is satisfactory if the correct cultivar is used. In some cases, seed borne viruses might be present in certain seed sources.

## Harvesting and Processing Seeds

### *Maturity and Ripening*

Each crop and plant species undergoes characteristic changes leading to seed ripening that must be known to establish the best time to harvest. A seed is ready to harvest when it can be removed from the plant without impairing germination and subsequent seed vigor. This is called **harvest maturity.** In many cases, a balance must be made between late and early harvest to obtain the maximum number of high-quality seeds. If harvesting is delayed too long, the fruit may **dehisce** ("split open" or "shatter"), drop to the ground, or be eaten or carried off by birds or animals. If the fruit is harvested too soon when the embryo is insufficiently developed, seeds are apt to be thin, light in weight, shriveled, poor in quality, and short-lived. Some seeds that are mechanically harvested (i.e., sweet corn) can be damaged if the seed moisture at harvest is too dry. Therefore, developing seeds are sampled often to determine their stage of maturity. Seed moisture percentages are used as an indicator of seed maturity to determine the proper harvest time. Early seed harvest may also be desirable for seeds of some species of woody plants that produce a hard seed covering in addition to a dormant embryo. If seeds become dry and the seed coats harden, the seeds may not germinate until the second spring , whereas they would have germinated the first spring if harvested early.

### *Harvesting and Handling Procedures*

Plants can be divided into three types for seed extraction, according to their fruit type:

1. Dry fruits that do not dehisce at maturity
2. Dry fruits that dehisce at maturity
3. Plants with fleshy fruits

#### Type 1: Dry Fruits That Do Not Dehisce at Maturity

Plants in this group have seed and fruit covers that adhere to each other at maturity. These are dry fruits that do not dehisce (open), and the seeds are not disseminated immediately upon maturity. This group includes most of the agricultural crops, such as corn, wheat, and other grains. Many of these have undergone considerable selection during domestication for ease of harvest and handling. This group also contains the nut crops like oak (*Quercus*), hazel (*Corylus*), and chestnut (*Castanea*).

Field-grown crops (cereals, grasses, corn) can be mechanically harvested using a combine, a machine that cuts and threshes the standing plant in a single operation (Fig. –4). Plants that tend to fall over or "lodge" are cut, piled, or windrowed for drying and curing. Low humidity is important during harvest. Rain damage results in seeds that show low vigor. The force required to dislodge seeds may result in mechanical damage, can reduce viability, and result in abnormal seedlings. Some of these injuries are internal and not noticeable, but they result in low viability after storage. Damage is most likely to occur if seed moisture is too high or low, or if the machinery is not properly adjusted. Usually less injury occurs if seeds are somewhat moist at harvest (i.e., up to 45 percent for corn).

Nut crops usually have an involucre covering (i.e., the cup of an acorn) that should be separated from the nut at harvest. Floatation is a common method for separating viable from non-viable seeds (Fig. –5, ). Floating seeds are more buoyant usually because of insect infestation and are discarded.

(a) (b) (c) (d)

*Figure –4 Corn seed is actually a fruit (caryopsis) and is an example of a crop with dry nondehiscent fruits. (a) Corn seed is harvested with a picker, leaving the kernels attached to the cob. Although corn used for grain is combined (harvested and shelled in one operation), corn for seed is usually not shelled until it is allowed to dry further to prevent mechanical injury. (b) Corn dehusker. (c) Dehusked corn cobs. (d) Shelled kernels (seeds) ready for storage.*

*Figure -5 Non-viable oak nuts (acorns) float in water, while viable seeds are more dense and sink.*

## Type 2: Dry Fruits That Dehisce at Maturity

These plants produce seeds from fruits that dehisce readily at maturity. This type includes seeds in follicles, pods, capsules, siliques, and cones. Crops of this group include many annual or biennial flowers (delphinium, pansy, petunia) and various vegetables (onion, cabbage, other cruciferous crops, and okra).

In most cases, these fruits must be harvested before they are fully mature, and then dried or cured before extraction. Consequently, some seeds will be underdeveloped and immature at the time of harvest. For example, in onion seed production there is a difference of up to 20 days between the opening of the first and last flowers on a plant. From a practical standpoint, onion seeds are harvested when the first fruits begin to open and seeds have turned black, which corresponds to about 50 days after flowers first open and begin shedding pollen.

In addition, many tree and shrub plants also have fruits that fall into this group and are handled with similar procedures. The steps for handling these types of seeds:

1. **Drying.** Plants are cut (sometimes by hand) or dry fruits may be windrowed in the field (Fig. -6), or placed on a canvas, tray, or screen (Fig. -7) to dry for 1 to 3 weeks. If there are only a few plants, they can be cut and hung upside down in a paper bag to dry. Some crops may need the benefit of forced air

drying units for quick dryings, especially in harvest areas with high humidity at the time of harvest (Fig. -7).

2. **Extraction.** Commercial seeds may be harvested and extracted in a single operation (Fig. -8) with a combine or dried fruits may be passed through threshing machines that extract seeds by beating, flailing, or rolling dry fruit followed by separation of seeds from fruit parts, dirt, and other debris (Fig. -9, ). Seeds from small seed lots are extracted by hand.
3. **Seed Conditioning (Cleaning).** Further cleaning may be required to eliminate all dirt, debris, weed, and other crop seeds. Commercial seed conditioning utilizes various kinds of specialized equipment, such as screens of different sizes (Fig. -10, ), seed shape (Fig. -11, ), air lifters (Fig. -12a and b), and gravity separators (Fig. 6-12c and d). The basis for these types of separation is that there are differences in sizes, shapes, and densities between good seed, poor seed, and other debris.

(a) (b)

*Figure -6 Cole crop (Brassica) seed production is also from a dry dehiscent fruit. (a) Turnip at full harvest maturity. (b) Cabbage seed field mowed and windrowed. Windrowing is done before the fruit shatters, and windrowing allows additional maturation and drying before being combined with a windrow pickup unit. (c) Cole crop fruit is a silique, which is a dry, dehiscent fruit that opens along two suture lines, exposing the seeds attached to a papery septum.*

(a) (b) (c)

*Figure -7 Seeds with non-dehiscent and dehiscent fruits often require additional drying after harvest. (a) Portable field drying wagons alongside a permanent bin dryer used for drying prairie wildflower seeds. (b) Open wire screen racks used for air drying woody plant seeds. (c) Forced-air dryer.*

*Figure – 8,Purple coneflower (Echinacea) seed production is an example of crop requiring the dry seed harvesting method. It has a fruit that shatters at maturity. (a) Seed production field in full bloom. (b) Field at harvest maturity before heads shatter and release seeds. (c) Combine for harvesting and threshing seeds. (d) The combine must be calibrated for cutting height and maximum seed retention. (e) The reel rotates and the paddles force plant stems into the (f) blades of the cutting bar.*

*Figure –9 Sandersonia seed removal from a dry dehiscent capsule. (a) Hand-cut fruiting stems are cut and windrowed under protective cover for additional drying. (b) Pods are passed through a threshing machine to remove seeds. (c) The threshing cylinder with a rasp-bar is the most common thresher. (d and e) Proper threshing captures up to 90 percent of the available seeds, but additional conditioning is usually needed to remove fruit debris.*

(a) (b) (c)

*Figure –10 Seed conditioning based on seed size and shape. (a) Hand screens manually sift seeds from plant debris; (b) Mechanical cleaner and seed sizing units use aspiration (air movement) combined with screens of various shapes and sizes to remove seed debris and separate seeds into various size classes. (c) Close up of screens in a scalper unit that separates good seed from plant debris and other unwanted material.*

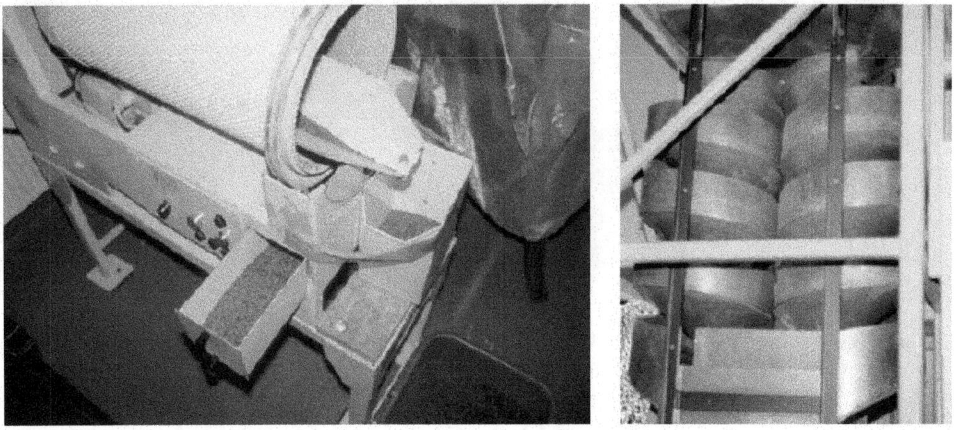

(a) (b)

*Figure –11 Seed conditioning based on seed shape. (a) An indent cylinder that separates seeds based on seed size (length). (b) A spiral separator uses gravity and centripetal force to separate round from flat seeds. Round seeds move faster down the separator. These are useful for cole crop seeds like cabbage and broccoli.*

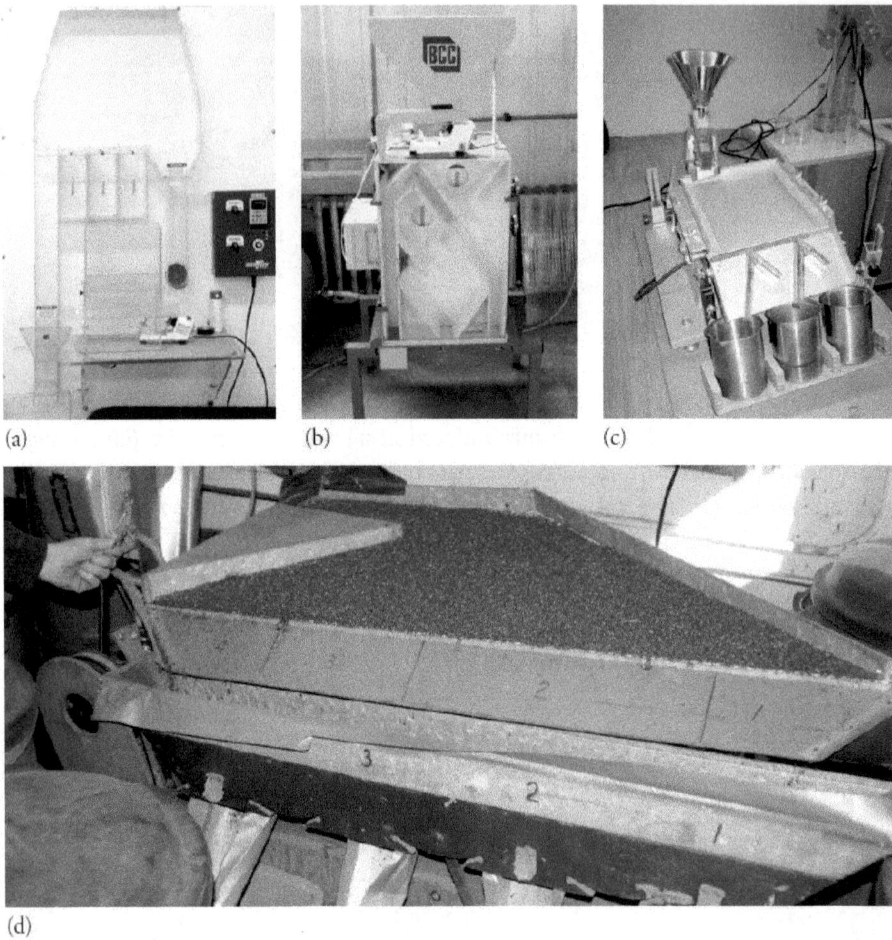


*Figure 6–12 Seed conditioning based on seed density. (a) The wall mounted air separator uses a vacuum to lift seeds. Seeds are separated from lighter plant debris. (b) Standalone movable air separator. (c and d) Gravity tables have a tilted platform that uses vibration or air flow to separate seeds. Denser seeds walk toward the higher edge of the platform. Both types of units can be used to upgrade seed lots by directing seeds into bins based on density (weight).*

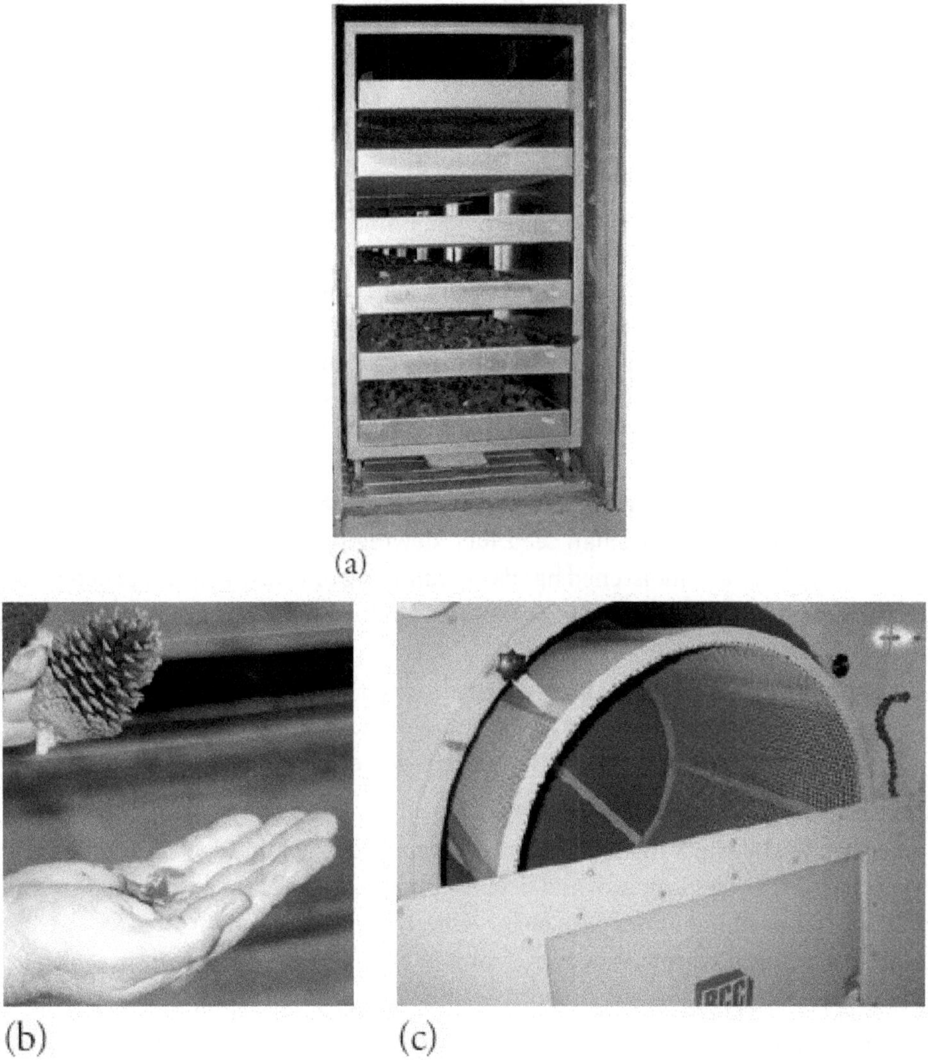


*Figure –13 Seed extraction and conditioning in pines (a) Drying oven used to release seeds from pine cones. (b) Winged seeds extracted from the cones. (c) Seeds are tumbled to remove the wing attached to the seeds.*

**Conifer cones** also fit in this category of dry dehiscent fruits, but their cones require special procedures :

1. **Drying.** Cones of some species will open if dried in open air for 2 to 12 weeks (Fig. –13). Others must be force-dried at higher temperatures in special heating kilns. Under such conditions, cones will open within several hours or, at most, 2 days. The temperature of artificial drying should be 46 to 60°C

(115 to 140°F), depending upon the species, although a few require even higher temperatures.

For example, Jack pine (*Pinus banksiana*) and red pine (*P. resinosa*) need high temperatures (77°C to 170°F) for 5 to 6 hours. Caution must be used with high temperatures, because overexposure will damage seeds. After the cones have been dried, the scales open, exposing the seeds.

2. **Extraction.** Seeds should be removed immediately upon drying, since cones may close without releasing the seeds. Cones can be shaken by tumbling or raking to dislodge seeds. A revolving wire tumbler or a metal drum is used when large numbers of seeds are to be extracted.
3. **Dewinging.** Conifer seeds have wings that are removed except in species whose seed coats are easily injured, such as incense cedar (*Calocedrus*). Fir (*Abies*) seeds are easily injured, but wings can be removed if the operation is done gently.

   Redwood (*Sequoia* and *Sequoiadendron*) seeds have wings that are inseparable from the seed. For small seed lots, dewinging can be done by rubbing the seeds between moistened hands or trampling or beating seeds packed loosely in sacks. For larger lots of seeds, special dewinging machines are used (Fig. –13c).
4. **Cleaning.** Seeds are cleaned after extraction to remove wings and other light chaff. As a final step, separation of heavy, filled seed from light seed is accomplished by gravity or pneumatic separators.

## Type 3: Plants with Fleshy Fruits

Plants with fleshy fruits include important fruit and vegetable species used for food such as berries, pomes (apples), and drupes (plums), as well as many related tree and shrub species used in landscaping or forestry. In general, fleshy fruits are easiest to handle if ripe or overripe. However, fruits in the wild are subject to predation by birds.

For **extraction** of small seed lots, fruits may be cut open and seeds scooped out, treaded in tubs, rubbed through screens, or washed with water from a high-pressure spray machine in a wire basket (Fig. –14). Another device that removes seeds from small-seeded fleshy fruits is an electric mixer or **blender** (Fig. –15). To avoid injuring seeds, the metal blade of the blender can be replaced with a piece of rubber or Tygon tubing. It is fastened at right angles to the revolving axis of the machine. A mixture of fruits and water is placed in the mixer and stirred for about 2 minutes. When the pulp has separated from the seed, the pulp is removed by flotation.

This procedure is satisfactory for fruits of serviceberry (*Amelanchier*), barberry (*Berberis*), hawthorn (*Crataegus*), strawberry (*Fragaria*), huckleberry (*Gaylussacia*), juniper (*Juniperus*), rose (*Rosa*), and others.

For larger lots, separation is by maceration, fermentation, mechanical means, or washing through screens. The basic procedures include:

1. **Maceration.** Vegetable crops such as tomato, pepper, eggplant, and various cucurbits are produced in commercial fields and may utilize special macerating machinery as a first step in seed extraction.

   Cucumber and other vine crops, for example, are handled with specially developed macerating machines (Fig. –16, ). Maceration crushes the fruits and mixes the pulverized mass with water that is diverted into a tank releasing the seeds, but additional handling is often required to separate seeds from the macerated pieces of fruit.

*Figure –14 Small seed lots of small, fleshy seeds can have the fruit pulp removed by rubbing fruits against a screen and washing away the pulp.*

2. **Fermentation.** Macerated fruits can be placed in large barrels or vats and allowed to ferment for up to 4 days at about 21°C (70°F), with occasional stirring.

Figure –15 A method for small batch extraction of seeds from fleshy fruits uses a blender (a) or food processor retrofitted with a rubber or plastic impeller for maceration followed by floatation (b and c) to remove seeds from the pulp. (d) Commercial macerators (i.e., Dybvig) use the same principles of water and flailing impellers to extract seeds. They work well for fruit crops like cherry, peach, and plum.

(a) (b) (c) (d) (e) (f)

*Figure-16.Watermelon is an example of a crop that requires the wet seed harvesting method for seed extraction from a fleshy fruit. (a) Field ready for harvest. Withholding water knocks down the vines prior to harvest. (b) Custom seed harvester for large fleshy fruit. (c) Fruit is crushed and the pulp is separated from the seeds. (d) Seeds with a small amount of adhering pulp. (e) A washing unit provides final separation of pulp and seeds. (f) Large rotating dryers reduce seed moisture to its storage level.*

If the process is continued too long, sprouting of the seeds may result. Higher temperature during fermentation shortens the required time. As the pulp separates from seeds, heavy, sound seeds sink to the bottom of the vat, and the pulp remains at the surface.

Following extraction, the seeds are washed and dried either in the sun or in dehydrators. Additional cleaning is sometimes necessary to remove dried pieces of pulp and other materials. Extraction by fermentation is particularly desirable for tomato seed, because it can help control bacterial canker.

3. **Chemical Treatment.** Alternatives to fermentation are various chemical treatments. The advantage of chemical treatments is that it takes less time (less than 24 hours) to separate seeds from macerated pulp.

   Like fermentation, overexposure to the chemical can reduce seed quality. Chemical treatments include acid treatment for tomato seed extraction , and digestive enzymes—like pectinase used in orange seed extraction—for understock production.

4. **Flotation.** Another alternative to separate seeds from fleshy fruits is floatation, which involves placing seeds and pulp in water so that heavy, sound seeds sink to the bottom and the lighter pulp, empty seeds, and other extraneous materials float to the top. This procedure can also be used to remove lightweight, unfilled seeds and other materials from dry fruits, such as acorn fruits infested with weevils, but sometimes both good and bad seeds will float. Small berries of some species, such as *Cotoneaster,* juniper (*Juniperus*), and *Viburnum,* are somewhat difficult to process because of small size and the difficulty in separating the seeds from the pulp. One way to handle such seeds is to crush the berries with a rolling pin, soak them in water for several days, and then remove the pulp by flotation.

(a) (b)

*Figure –17 Various drying units for seeds. (a) A spinning centripetal dryer. (b) A large rotating forced air dryer.*

After seeds are thoroughly washed to remove fleshy remnants, they are dried (Fig. –17), except seeds of recalcitrant species that must not be allowed to dry out. If left in bulk for even a few hours, seeds that have more than 20 percent moisture will heat; this impairs viability. Drying may either occur naturally in open air if the humidity is low or artificially with heat or other devices. Drying temperatures should not exceed 43°C (110°F); if the seeds are quite wet, 32°C (90°F) is better. Drying too quickly can cause seeds to shrink and crack, and can sometimes produce hard seed coats. The minimum safe moisture content for storage of most orthodox seeds differs by species but is usually in the range of 4 percent to 15 percent.

## *Seed Testing*

In the United States, state laws regulate the shipment and sale of agricultural and vegetable seeds within each state. Seeds entering interstate commerce or those sent from abroad are subject to the **Federal Seed Act,** adopted in 1939. Such regulations require the shipper to use labeling (Fig. –18) of commercially produced seeds that includes:

1. Name and cultivar
2. Origin
3. Germination percentage
4. Percentage of pure seed, other crop seed, weed seed, and inert material

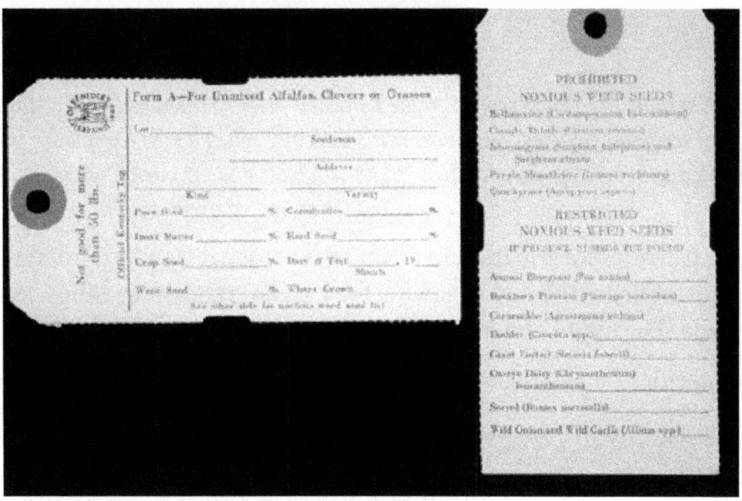


*Figure –18 State and federal seed laws require testing seed lots prior to sale. Information for a seed lot includes standard germination percentage according to accepted seed-testing rules, purity of the seed lot (percentage of seeds that are the desired crop and its trueness to type), percentage of weed seeds, and the amount of noxious weed seeds in the seed lot. Noxious weeds are designated as weeds that are particularly undesirable, and tolerances may differ for a crop or region of the country.*

Regulations set minimum standards of quality, germination percentage, and freedom from weed seeds. Special attention must be paid to designated **noxious weeds** for a particular growing region. Laws in some states and in most countries regulate shipment and the sale of tree Seed , but there are no federal laws governing the tree seed trade.

*Figure -19 A sample from each seed lot must be tested prior to saleusually by a state-certified seed lab. The seed analyst uses a seed sorter to randomly select a seed sample for testing from the submitted seed lot. A portion of the seed lot will be tested for purity, while an additional subsample will be evaluated for standard germination.*

Seed testing provides information in order to meet legal standards, determines seed quality and establishes the rate of sowing for a given stand of seedlings. It is desirable to retest seeds that have been in storage for a prolonged period.

International rules for testing seeds are published by the **International Seed**

## Testing Association *(www.seedtest.org).*

A high-quality seed lot is a function of the following characteristics that are routinely tested by seed companies or private and state seed labs :

1. Germination (viability)
2. Purity
3. Vigor
4. Seed health
5. Noxious weed seed contamination

## Sampling for Seed Testing

The first step in seed testing is to obtain a uniform sample that represents the entire lot under consideration (Fig. –19). Equally sized (usually measured by weight) primary samples are taken from evenly distributed parts of the seed lot, such as a sample from each of several sacks in lots of less than five sacks or from every fifth sack with larger lots. The seed samples are thoroughly mixed to make a composite sample. A representative portion is used as a submitted sample for testing. This sample is further divided into smaller lots to produce a working sample (i.e., the sample upon which the test is actually to be run). The amount of seed required for the working sample varies with the kind of seed and is specified in the Rules for Seed Testing.

## Viability Determination

Viability can be determined by several tests, the **standard germination, excised embryo,** and **tetrazolium** tests being the most important. Standard Germination Tests In the standard germination test, germination percentage is determined by the percent of **normal seedlings** produced by pure seeds (the kind under consideration).

To produce a good test, it is desirable to use at least 400 seeds picked at random and divided into lots of 100 each. If any two of these lots differ by more than 10 percent, a retest should be carried out. Otherwise, the average of the four tests becomes the official germination percentage.

Seeds are placed under optimum environmental conditions of light and temperature to induce germination. The conditions required to meet legal standards are specified in the rules for seed testing, which may include type of test, environmental conditions, and length of test.

Various techniques are used for germinating seeds in seed-testing laboratories. Small seeds are placed on plastic germination trays or in Petri dishes (Fig. –20). The most common substrate used by commercial seed technology labs for germination tests are blue blotter or washed paper towels, available from commercial suppliers. These products ensure uniformity and reproducibility in their tests. Containers are placed in germinators in which temperature, moisture, and light are controlled according to the established standard germination rules. To discourage the growth of microorganisms, all materials and equipment should be kept scrupulously clean, sterilized when possible, and the water amount carefully regulated.

The **rolled towel test** (Fig. –21a, b, and c, ) is commonly used for testing large seeds like cereal grains. Several layers of moist paper toweling, about 2.8 by 3.6 cm (11 by 14 in) in size, are folded over the seeds and then rolled into cylinders and placed vertically in a germinator.

A germination test usually runs from 1 to 4 weeks but could continue up to 3 months for some slowgerminating tree seeds with dormancy. Usually a first count is taken at 1 week and germinated seeds are discarded with a final count taken later. At the end of the test, seeds are divided into (a) normal seedlings, (b) hard seeds, (c) dormant seeds, (d) abnormal seedlings, and (e) dead or decaying seeds. A normal seedling should have a well-developed root and shoot, although the criterion for a "normal seedling" varies with different kinds of seeds (Fig. –21d). "Abnormal seedlings" can be the result of age of seed or poor storage conditions; insect, disease, or mechanical injury; frost damage; or mineral deficiencies. Any non-germinated seeds should be examined to determine the possible reason. "Hard seeds" have not absorbed water. Dormant seeds are those that are firm, swollen, and free from molds but do not germinate.

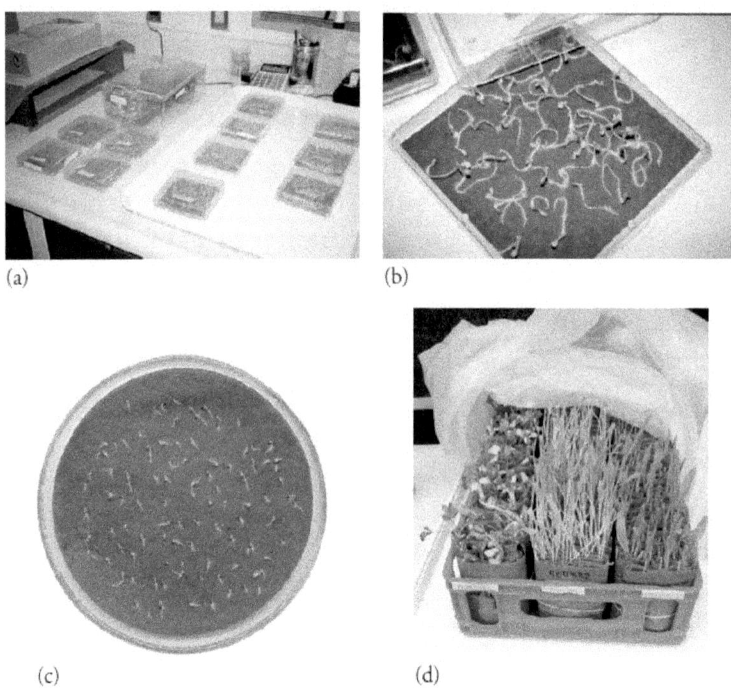


*Figure –20 A standard germination test is required for seed lots prior to sale. The two most common test procedures include the (a, b, and c) Petri dish and (d) rolled towel tests. The tests and the procedures used for standard germination are detailed in accepted publications like the rules for testing seeds from the Association of Official Seed Analysts. Included in these rules will be the preferred test (i.e., Petri dish or rolled towel); the environment for the test (i.e., 20/30, this indicates daily cycles of 16 hours at 20°C followed by 30°C for 8 hours); whether light is required during the test; any seed pretreatments for dormant seeds (e.g., treatment with gibberellin or potassium nitrate); and the number of days for the first and last evaluation (counts).*

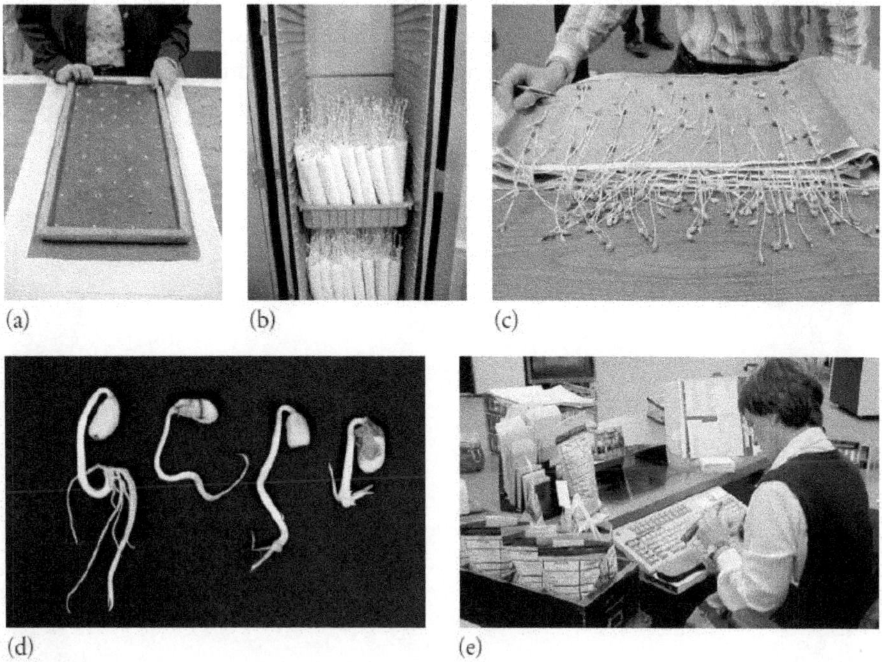

*Figure –21 Commercial seed labs process a large number of seed samples. They must keep accurate records of each seed lot and must be efficient to process samples in a timely manner while maintaining high reproducibility from seed lot to seed lot. (a) A seed analyst uses a template board to place a standard number of seeds in precise locations on the germination paper for the rolled towel or Petri dish tests. (b) Rolled towels are held upright in the growth chamber. (c) After the number of days indicated in the testing rules, the seed analyst counts the number of normal seedlings. (d) The seed analyst must determine if a seedling is normal and can be counted as germinated. These seedlings are "abnormal" because either the shoot or root has not developed normally after the final count for this seed test. (e) Results are recorded in a computer database.*

Under seed-testing rules, certain environmental requirements to overcome dormancy may be specified routinely for many kinds of seeds. These may include chilling stratification or hormone treatment with gibberellins or potassium nitrate.

**Excised-Embryo Test** The excised-embryo test is used to test seed viability of woody shrubs and trees whose dormant embryos require long treatment periods to relieve dormancy before true germination will take place. In this test, the embryo is excised from seeds that are soaked for 1 to 4 days and germinated following standard germination conditions (see Fig. –22).

The excision must be done carefully to avoid injury to the embryo. Any hard, stony seed coverings, such as the endocarp of stone fruit seeds, must be removed first.

The moistened seed coats are cut with a sharp scalpel, razor blade, or knife, under clean but nonsterile conditions with sterilized instruments. The embryo is carefully removed. If a large endosperm is present, the seed coats may be slit and the seeds covered with water, and after about a half-hour, the embryo may float out or be easily removed.

*Figure –22 The excised-embryo test is a quick evaluation method used for dormant seed. Eastern redbud (Cercis) seeds require at least four months of moist chilling to satisfy dormancy and another 2 weeks for a standard germination test. In comparison, isolated embryos removed from the seed coverings will germinate in 5 days.*

Tetrazolium Test The **tetrazolium test** is a biochemical test for viability determined by the red color appearing when seeds are soaked in a 2,3,5-triphenyltetrazolium chloride (TTC) solution (Fig. –23).

Living tissue changes the TTC to an insoluble red compound (chemically known as formazan); in nonliving tissue the TTC remains uncolored. The test is positive in the presence of dehydrogenase enzymes involved in respiration. This test was developed in Germany by Lakon , who referred to it as a **topographical test** since loss in embryo viability begins to appear at the extremity of the radicle, epicotyl, and cotyledon tips. The reaction takes place equally well in dormant and nondormant seed. Results can usually be obtained within 24 hours. The TTC solution deteriorates with exposure to light but will remain in good condition for several months if stored in a dark bottle. The solution should be discarded if it becomes yellow. A 0.1 to 1.0 percent concentration is commonly used. The pH should be 6 or 7. In the hands of a skilled technologist, this test can be used for seed-quality evaluation and as a tool in seed research.

*Figure –23 Tetrazolium chloride (TZ) is used to test seed viability. Portions of the embryo will stain red (an indication of respiration) if they are viable. The seed analyst must determine if vital portions of the embryo are living, which would indicate positive germination potential. (a and b) A positive TZ corn seed test showing that the embryo and scutellum are viable while the white endosperm is non-living at maturity. (c and d) A poor TZ test in gasplant (Dictamnus). White embryos are non-viable and the embryo (d) although generally red-stained would probably be abnormal because the shoot area (arrow) did not stain.*

X-ray analysis of seeds can be used as a rapid test for seed soundness. X-ray photographs do not normally measure seed viability but provide an examination of the inner structure for mechanical disturbance, absence of vital tissues, such as embryo or endosperm, insect infestation, cracked or broken seed coats, and shrinkage of interior tissues (Fig. -24).

Standard X-ray equipment is used to assess seeds. Dry seeds are exposed for 1/2 to 3 minutes at 15- to 20-kilovolt tube potential. Seed with dimensions less than 2 mm are too small to show details. Since X-rays do not injure the seed, further tests for viability can be conducted on the same batch. Prototype machines that provide fast, automatic, online sorting have been proposed. These procedures have the potential to remove nonviable seeds as well as seeds with morphological characteristics that are linked to poor vigor.

(a) (b)

*Figure -24 Examples of the X-ray tests for the 1999 (a) and 2005 (b) harvests of Gaura biennis capsules. Note the number of filled and empty (aborted) seeds in the capsules.*

## *Purity Determination*

**Purity** is the percentage by weight of the "pure seed" present in a sample. Purity determination requires a trained seed analyst, usually from a state or private seed lab. In India, the **Society of Commercial Seed Technologists** provides training and testing to certify **Registered Seed Technologists.**

There are two aspects to pure seed: a **physical** and a **genetic** component. Pure seed must be separated from other physical contaminants such as soil particles, plant debris, other inert material, and weed seeds (Fig. -25). Seed standards list tolerances for levels of pure seed in a sample. They usually are based on the seed type and seed class (i.e., Certified vs. Registered seed). References are available with detailed seed anatomy to help seed technologists to identify crop and weed seeds.

Special care must be taken to document the occurrence of noxious weeds in a sample. **Noxious weeds** are identified as being particularly bad weeds for a region of the country and can vary by state.

Occurrence of a single seed of some noxious weed species in a sample can render an entire seed lot unacceptable for public sale.

Purity testing also identifies the genetic purity of a seed lot. The seed analyst determines if the sample is the proper cultivar and identifies the percentage of seeds that are either other contaminating cultivars or inbreds in a hybrid seed lot. Genetic purity can be difficult to determine and relies on an assortment of tests that include field visits by regulatory personnel, seed color, seed and seedling morphology, chemical tests, isozyme (characteristic seed proteins) separation by electrophoresis, and DNA fingerprinting .

## *Vigor Testing*

Although state and federal seed laws currently require only purity and standard germination tests for seed lots, seed companies and many crop producers are performing **vigor** tests prior to sale or use. The Association of Official

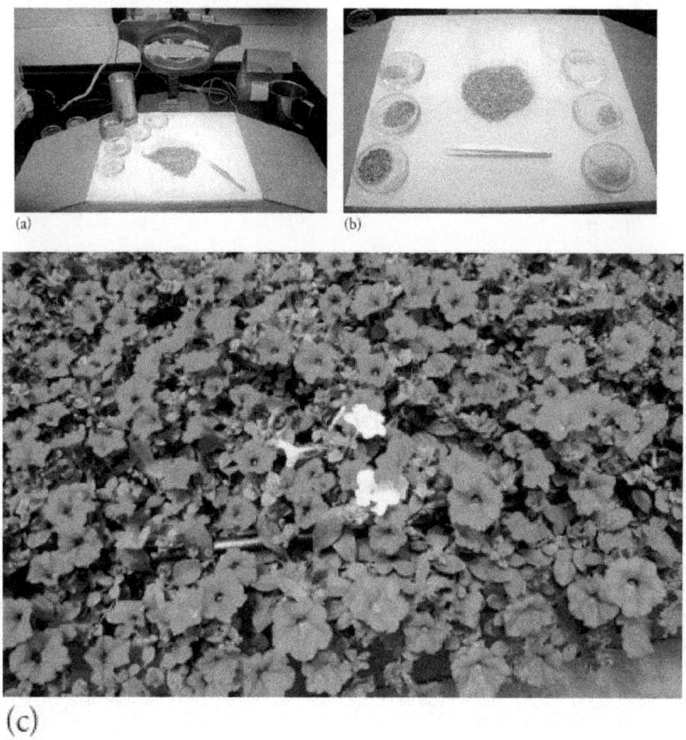


*Figure -25,Purity of seeds is determined by visual examination of individual seeds in a weighed seed sample taken from the larger lot in question. (a) Impurities may include other crop seed, weed seed, and inert, extraneous material. In this seed lot, several different types of impurities were discovered in the seed lot. (b) Each was placed in a small dish and will be weighed. (c) Purity is also evaluated in field or greenhouse trials. This petunia seed lot shows a percentage of white variants reducing its purity. White plants may be from self-pollinated plants from the female inbred parent that should have been removed during production.*

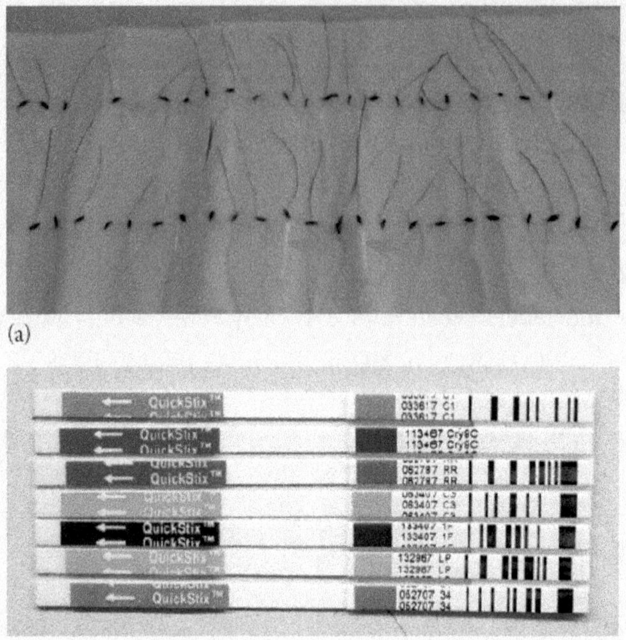


(a) (b)

*Figure –26 Biochemical and genetic tests for purity. (a) Fluorescence ryegrass tests. Under ultraviolet light, perennial ryegrass can be separated from annual ryegrass (red arrow) because of the characteristic white fluorescence of the annual ryegrass radicle. (b) A genetic strip test for the presence of genetically modified seeds. Each strip recognizes a different genetic modification using antibodies for traits such as insect and herbicide resistance. Similar to a pregnancy test, the white portion of the strip produces a single line for negative and a double line for a positive identification.*

Seed Analysts states that "seed vigor comprises those seed properties which determine the potential for rapid, uniform emergence, and development of normal seedlings under a wide range of field conditions." Standard germination tests do not always adequately predict seedling emergence under field conditions (Fig. –27). Seed vigor tests can provide a grower with additional information that can help predict germination where conditions may not be ideal. For many vegetable crops, there is a positive relationship between seed vigor and crop yields.

Various vigor tests have been developed and certain tests are applied to different species. Vigor tests include accelerated aging, controlled deterioration, cold test, cool test, electrolyte leakage, seedling growth rate, and seedling grow-out tests.

## Seed Health

Seed companies usually have the personnel and facilities to evaluate the health of a seed lot. Seed health comprises the occurrence of diseases, insects, or nematodes in the seed lot. Detection of these organisms requires specialized equipment and trained personnel. Seed health is integral to the performance of the seed lot. It has also become increasingly important as international trading agreements.

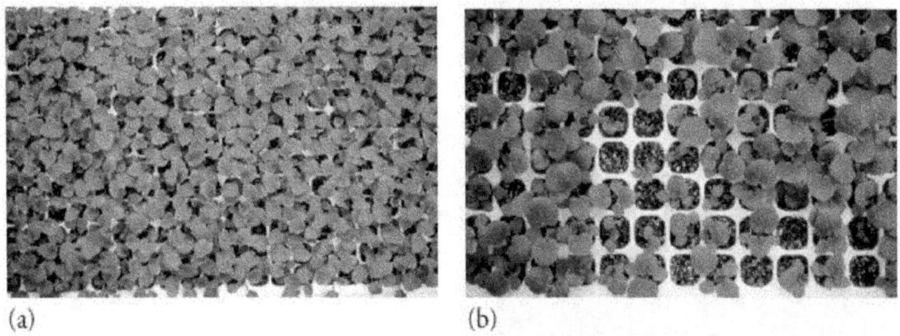

*Figure -27 Here is a good example of the impact of seed vigor on stand establishment. In both cases, all pansy seeds have germinated in each plug flat, but seedlings on the left (a) are all at the same stage of growth, while the plug flat on the right (b) has numerous seedlings that are less developed than the majority of seedlings.*

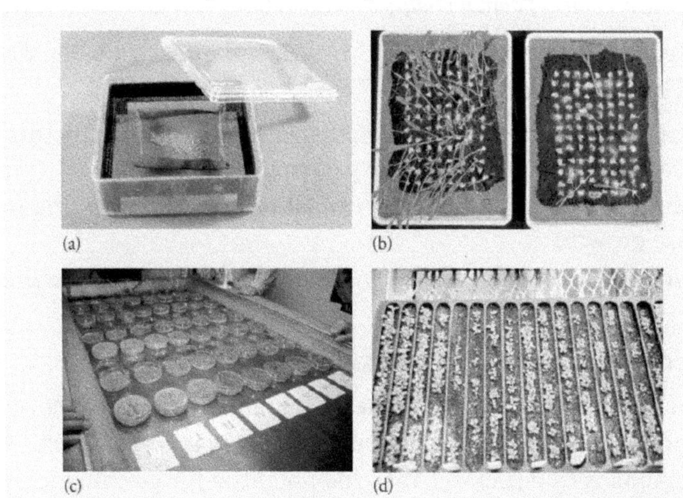

*Figure -28 Various seed vigor tests. (a) Impatiens seeds in accelerated aging boxes. The frame inside the box keeps seeds suspended above water or a solution of saturated salts. (b) Sweet corn seeds sprouting in the cold test. Seeds are placed on moist towels or Kimpack and covered with field soil. It is easy to see that the seed lot on the left has higher vigor (seedling emergence) compared with the seed lot on the right. (c) A thermal gradient table provides numerous temperatures to simultaneously test germination of a single seed lot, which is useful for determining seed vigor by evaluating germination at minimal and maximal temperatures. Breeders also use thermal gradient tables to evaluate a genotype's tendency for producing seed susceptible to thermodormancy (like lettuce). (d) For many horticultural crops, standard germination and seedling vigor is evaluated in a seedling grow-out test. The environment for this test is standard greenhouse conditions where the crop will be commercially grown.*

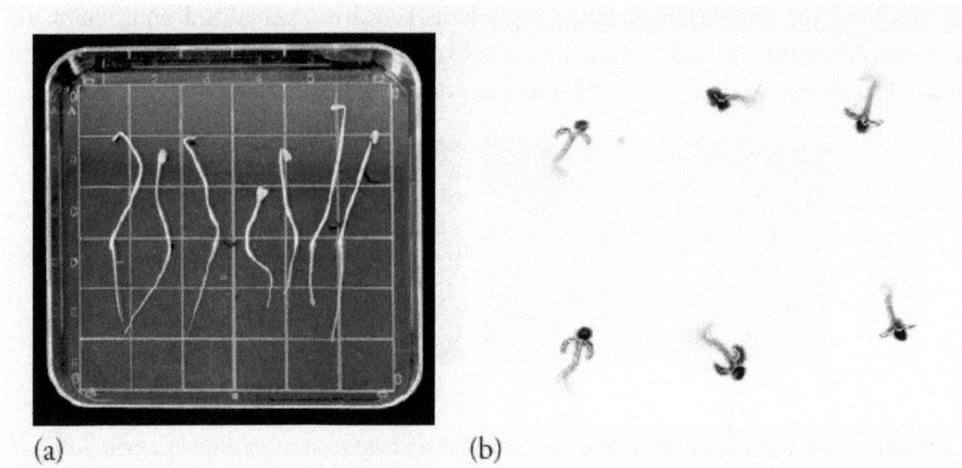

*Figure –29 (a) A slant-board test for lettuce. Seedling must be grown in an upright orientation to get straight seedlings. Radicle length is then measured by hand. (b) Computer-aided measurements of digital images of petunia from Petri dish germination.*

Specific procedures to standardize seed health tests are available. Three types of tests for seed health include:

1. **Visual evaluation** of a seed sample for characteristic structures like spores or sclerotia of pathogens, or the presence of insects.
2. **Incubation of seed** on moist germination paper or agar and inspection for disease growth.
3. **Biochemical tests,** such as ELISA tests, which detect the presence of specific disease organisms.

## Seed treatments to improve Germination

Pre sowing seed treatments has become a common practice in the seed industry. Seed treatments may be applied by seed producers or on the farm. The objective of seed treatments is to either enhance the potential for germination and seedling emergence or to help mechanical seed sowing.

Types of seed treatments include:

1. Seed protectants
2. Germination enhancement
3. Inoculation with microorganisms (nitrogen-fixing bacteria)
4. Coatings to help mechanical sowing Facilities that treat seeds must consider the following aspects for quality seed treatment :
   1. Seeds must be treated **uniformly.**
   2. The material must continue to **adhere to the surface** of the seed during sowing.

3. The treatment should **not reduce seed quality.** Any physical damage due to high temperature or mechanical injury must be minimized and monitored by seed testing.
4. The treatment should be **safely applied** and allow for safe handling by the seed consumer.
5. Treatments to **help mechanical sowing** must produce a **uniform size** and **shape** for each seed.
6. All seeds treated with a pesticide must be **colored** to avoid accidental ingestion by humans or animals. Color can also enhance the appearance of the seed.

Modern seed treatments require specialized equipment and facilities. The equipment varies depending on the type of seed treatment.

Historically, the first seed treatment incorporated pesticides in simple powders. These are still used today, especially for on-site farm application, because they require the least specialized equipment. However, powders and the dust from them present a problem for safe handling. Most commercial treatment of seeds is from **liquid slurries.** These are preferred because they treat seeds more uniformly, are safer to apply and handle, and are relatively cheap.

Recently, **polymer film coatings** have become a popular seed treatment because the pesticide can be incorporated into the polymer that is applied in a thin, uniform coat or film. The advantages of film coatings are the ability to incorporate chemical or biological materials into the coating for safe handling (this material does not rub off when handled), uniform coating size, and an attractive appearance. The cost has been prohibitive for general use with many large-volume agronomic crops, but film-coated seeds have become more widely available on high-value flower and vegetable seed.

## Seed Protectants

**Seed protectants** can be grouped as

1. **Chemical treatments** against pathogens, insects, and animals.
2. **Heat treatment** against pathogens and insects.
3. **Inoculation** with **beneficial microbes** against harmful fungi.
4. **Safners,** to reduce herbicide injury.

**Chemical Treatment** A seed stores food reserves to provide energy and carbon for seedling growth, which makes seeds a primary food source for humankind. However, insects, pathogens, and animals also target seeds as a food source. Strategies to protect seeds probably date to man's earliest use of seeds as a food crop. Chemical treatments for seeds can be seen in the 1800s with the use of copper sulfate against a variety of cereal diseases.

In the 1900s, mercury compounds were very effective against seed and seedling pathogens. These were banned in most parts of the world in the 1980s because of health risks. The 1940s and 1950s saw the introduction of the first broad-spectrum fungicides (like captan and thiram), starting the modern use of seed protectants for diseases.

The most common and important seed treatments are the chemical and physical treatments against seed-borne pathogens and insects. It is important to understand that these treatments will not improve germination in seeds with a genetically low potential for germination or in mechanically injured seeds. These treatments are especially beneficial where germination is delayed due to poor environmental conditions such as excessive water in the field, or cool soils.

Under these conditions, seed leakage stimulates fungal spore germination and growth. A chemical seed treatment can protect the seed until the seedling emerges.

Seed treatment may be designed to protect seed from soil-borne pathogens, disinfest the seed from pathogens on the seed surface, or eliminate pathogens inside the seed. Chemical seed protectants can be applied as **powders, liquids, slurries,** or incorporated into a pellet or **film coating.**

**Biocontrol** Although chemical treatments dominate industry seed treatments, the novel use of treating seeds with **beneficial microbes** presents an interesting alternative to chemical treatments. Various biocontrol agents provide protection to seeds by producing antibiotic substances; decreasing competition for space and nutrients; and reducing parasitism. Common biocontrol agents include bacterial strains like *Eterobacter, Pseudomonas, Serratia,* and fungal strains like *Gliocladium* and *Trichoderma.*

**Heat Treatment (Thermotherapy)** High temperature to control seed-borne diseases has been in use since 1907. Dry seeds are immersed in hot water (49 to 57°C; 120 to 135°F) for 15 to 30 minutes, depending on the species. After treatment, the seeds are cooled and spread out in a thin layer to dry. To prevent injury to the seeds, temperature and timing must be regulated precisely; a seed protectant should subsequently be used, and old, weak seeds should not be treated. Hot water is effective for specific seed-borne diseases of vegetables and cereals, such as *Alternaria* blight in broccoli and onion, and loose smut of wheat and barley.

Microwave and UV radiation also can be used to disinfest seeds. Aerated steam is an alternate method that is less expensive, easier to manage, and less likely to injure seeds than hot water. Seeds are treated in special machines in which steam and air are mixed and drawn through the seed mass to rapidly (in about two minutes) raise the temperature of the seeds to the desired temperature. The treatment temperature and time vary with the organism to be controlled and the kind of seed. Usually the treatment is 30 minutes, but it may be as little as 10 or 15 minutes.

Temperatures range from 46 to 57°C (105 to 143°F). At the end of treatment, temperatures must be lowered rapidly to 32°C (88°F) by evaporative cooling until dry. Holding seeds in moisture-saturated air at room temperature for 1 to 3 days prior to the steam-air treatment will improve effectiveness.

Hot water is also used to kill insects in seeds. For example, oak (*Quercus*) seed is soaked in water at 49°C (120°F) for 30 minutes to eliminate weevils commonly found in acorns. As with heat treatments to eliminate disease, precise temperature and timing must be maintained or seeds will be damaged.

### Seed Coating

Seed coating uses the same technology and equipment used by the pharmaceutical industry to make medical pills. Seed coatings include **pelleted** and **film-coated** seeds.

**Pelleted Seeds** The objective of coating seeds as a pellet is to provide a **round, uniform shape** and **size** to small or unevenly shaped seeds in order to aid precision mechanical sowing (Fig. –30). Pelletized seeds are tumbled in a pan while inert powders (like clay or diatomaceous earth) and binders form around seeds to provide a uniform, round shape (Fig. –31). Recent advances in coating materials and processing using rotary coaters has allowed seed producers to produce thinner pellets (Fig. –30b). These are usually termed **encrusted** seeds for very thin coatings (1 to 5 times the seed size) or mini-pellets (10 to 25 times the seed size). Compare this with a traditional pellet that may be 50 to 100 times the seed size (Fig. –30c and d).

Encrusted seeds are similar to film-coated seeds but are less expensive to produce. Pellets can be distinguished by either "splitting" or "melting" when the coating is wetted, with many growers preferring the split-type pellets (Fig. –30e). Many ornamental flower seeds are commonly pelletized for precision sowing one seed per cell in a plug flat. An increasing number of direct-seeded vegetable crops are also being pelletized. It is common for lettuce seed sown in Florida and California to be pelletized to provide uniform spacing and sowing depth that reduces the need to hand-thin the crop.

**Polymer Film-Coated Seeds** Film coating (Fig. –32, ) uses a thin polymer film to cover the seed. Film coating only adds 1 to 5 percent to the weight of a seed compared with more than 1,000 percent for pelletized seed, but this can still aid in precision sowing by improving flowability. Fungicides and beneficial microbes can be added to both pellets and film coatings and is the major benefit to film coating. Novel films are being employed that allow seeds to imbibe only when the soil temperature has warmed to prevent imbibitional chilling injury in sensitive plants.

## Germination Enhancement

Commercial practices that provide germination enhancement are **seed sizing, priming,** and **pregermination.**

**Seed Sizing** Seed lots sold as "elite" seeds have been sized to provide larger seed. In addition, seed sizing eliminates lightweight and cracked seeds (Fig. -33). This can provide seeds with a higher potential for germination viability and vigor. Elite seeds also may be the seeds selected by seed companies to be further enhanced by seed priming.

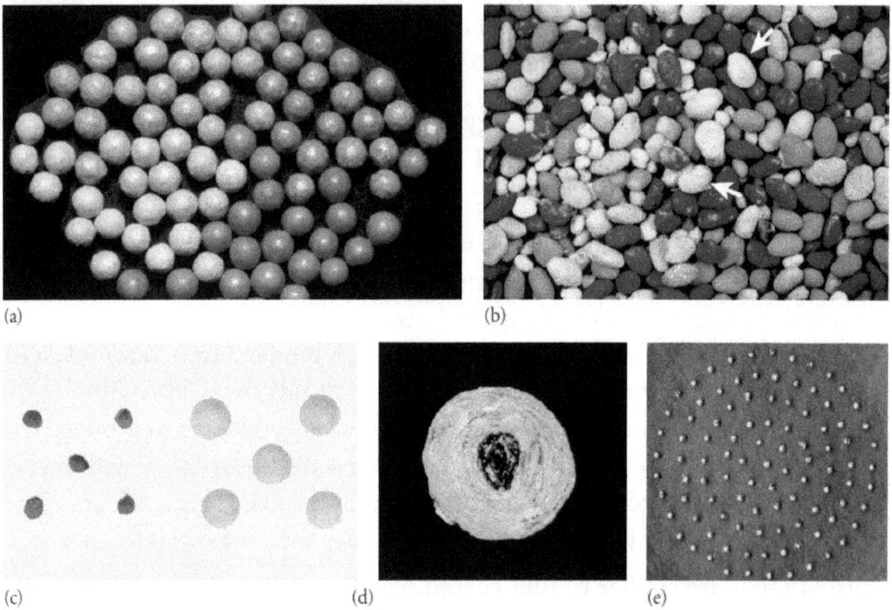

*Figure -30 Seed Pellets (a) Pelleted seeds showing the uniformly round shape to help in mechanical sowing. Colors may indicate seed differences (primed vs. untreated) or just be cosmetic. (b) A collection of encrusted pasture legume seeds. Notice how the seed shape is still evident with the lighter pelleti coating; the arrows indicate non-encrusted seeds. (c and d) Seed pelleting adds considerable size to a seed as well as a uniform, round shape. (c) On the left are raw seeds versus pelleted seeds on the right. (d) A cross-section of a pelleted seed showing how the coating (light blue) adds significant volume to the seed. (e) Pellets showing the split-coat habit as it hydrates. Splitting allows easy penetration by the radicle of the germinating seed.*

**Seed priming** A controlled hydration seed treatment that induces faster, more uniform germination. This effect is most noticeable when seeds are sown in less-than-favorable environments. Primed seeds are most often used in greenhouse bedding plant production to shorten the time to produce seedling plugs, and in crops like pansy and lettuce to avoid reduced germination due to high temperature.

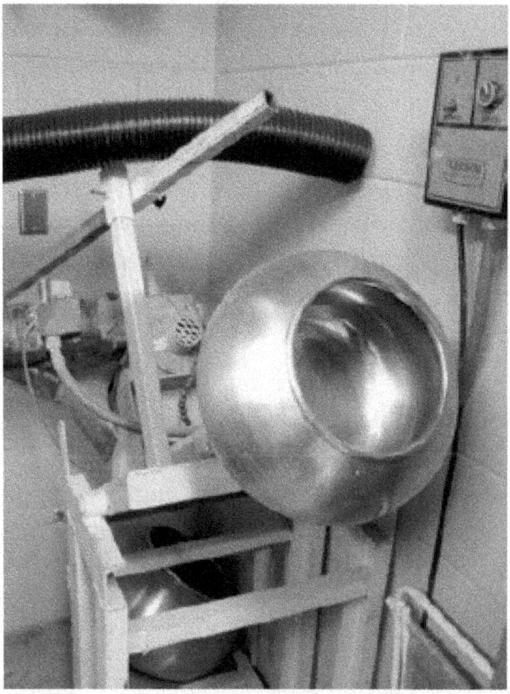

*Figure –31 Pan type seed coater for pelletizing seeds. Seeds tumble in this seed coating machine while layers of a bulking material and binder build the pellet around the seed.*

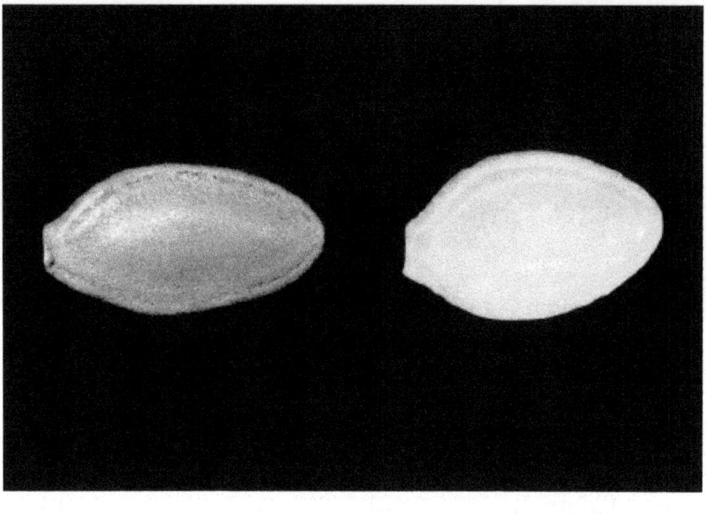

*Figure –32 Film coating is used to improve flowability of seeds during planting and as a carrier for pesticides. Several examples of film coating on corn seed. Seeds on the left are untreated.*

(Radicle Emergence). After being hydrated for an extended time, seeds are dried back to near the original dry weight. These seeds can be handled as normal raw seeds or pelleted prior to sowing. Growth substances or biologicals (termed *biopriming* ;) also can be included in the priming solution for added seed enhancement.

Primed seeds will usually show higher seed vigor compared with raw seeds.. Priming can provide faster, more uniform seedling emergence for field and greenhouse crops, especially when environmental conditions for germination are not ideal. The grower must weigh the additional cost of primed seed with this potential for improved seedling emergence. It is common to prime crops like lettuce and pansy to overcome problems of reduced germination due to conditions of high temperature.

**Pregermination** The goal of each grower is to establish a "stand" (seedling emergence) of 100 percent , which means a plant at each appropriate field spacing or greenhouse plug cell. This can be accomplished by using transplants or sowing more seeds than are required and thinning seedlings to the appropriate spacing. An additional treatment to improve stand establishment is pregermination of seeds. In concept, pregermination can take place under optimum conditions and any seeds showing radicle emergence are sown, providing near 100 percent stand. Two types of pregermination sowing techniques have been used:

1. **Fluid drilling** to sow germinated seeds in a gel to protect emerged radicles.
2. **Pregerminated** seeds that use a technique to dry seeds after the radicle emerges prior to sowing.

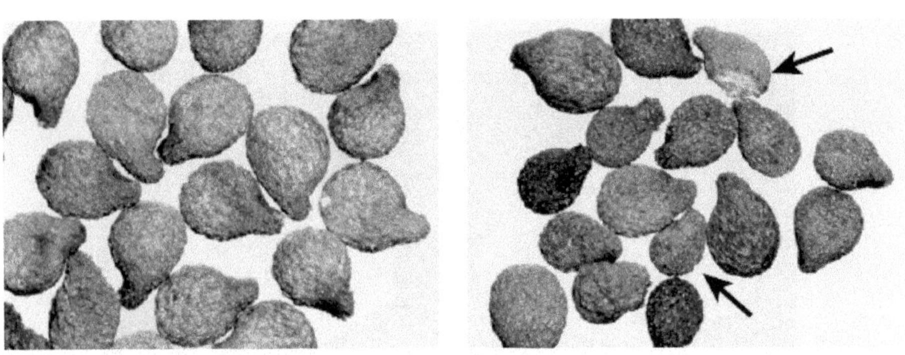

(a) (b)

*Figure -33 (a) "Elite" or enhanced seeds have additional seed conditioning to remove any broken seeds and have been sized to give larger, more uniform seeds. (b) Notice the broken and small seeds (arrow) in the seed lot on the right.*

Fluid Drilling. **Fluid drilling** is a system involving the treatment and pregermination of seeds followed by their sowing suspended in a gel. Seeds are pregerminated under conditions of aeration, light, and optimum temperatures for the species (Fig.

-22). Among the procedures that can be used are (a) germinating seeds in trays on absorbent blotters covered with paper, or (b) placing seeds in water in glass jars or plastic columns through which air is continuously bubbled and fresh water continuously supplied. Growth regulators, fungicides, and other chemicals can potentially be incorporated into the system. Chilling (10°C, 50°F) of thermodormant celery seeds for 14 days has produced short, uniform radicle emergence without injury.

Pregerminated seeds of various vegetables have been stored for 7 to 15 days at temperatures of 1 to 5°C (34 to 41°F) in air or aerated water. Separating out germinated seeds by density separation has improved the uniformity and increased overall stand.

Various kinds of gels are commercially available. Among the materials used are sodium alginate, hydrolyzed starch-polyacrylonitrile, guar gum, synthetic clay, and others. Special machines are needed to deposit the seeds and gel into the seed bed.

Pregerminated Seeds. **Pregerminated seeds** were introduced commercially in 1995 for bedding plant species (impatiens), but L. H. Bailey introduced the concept as early as 1897.

A quote from Bailey's *The Nursery Book* demonstrates that "new" is truly a relative term as he describes **"regermination."** "It is a common statement that seeds can never revive if allowed to become thoroughly dry after they have begun to sprout. This is an error. Wheat, oats, buckwheat, maize, pea, onion, radish, and other seeds have been experimented upon in this direction, and they are found to regerminate readily, even if allowed to become thoroughly dry and brittle after sprouting is well progressed. They will even regerminate several times."

Pregermination involves germination of seeds under controlled conditions to synchronize germination in order to induce the radicle to emerge about one-sixteenth of an inch. Germinated seeds are separated from nongerminated seeds, and then seeds are dried slowly to near their original dry weight. The advantages of using pregerminated seeds include production of 95 percent or better usable seedlings; fast, uniform germination; and because the seeds are dry, mechanical seeders can be used to sow them. The disadvantages of using pregerminated seeds are increased cost (up to 25 percent), seeds have a shorter shelf life (around 35 days at 5°C or 40°F), and growers must have optimized seedling growing conditions to take advantage of the benefits of pregermination.

## Seed Storage

Seeds are usually stored for varying lengths of time after harvest. **Viability** at the end of storage depends on (a) the initial viability at harvest, as determined by factors of production and methods of handling; and (b) the rate at which deterioration takes place. This rate of physiological change, or aging , varies with the kind of seed and the environmental conditions of storage, primarily temperature, and humidity.

## Seed Longevity

Plant species can be separated as **recalcitrant** or **orthodox** seeds based on their genetic potential to tolerate storage.

**Recalcitrant or Short-Lived Seeds Recalcitrant** seeds do not tolerate significant drying after seed development. Most recalcitrant seeds cannot tolerate seed moistures below 25 percent, and some species are also sensitive to chilling temperatures. This group is represented by species whose seeds normally retain viability for as little as a few days, months, or at most a year following harvest. However, with proper handling and storage, seed longevity may be maintained for significant periods..

The group includes:

1. Certain spring-ripening, temperate-zone trees such as poplar (*Populus*), maple (*Acer*) species, willow (*Salix*), and elm (*Ulmus*). Their seeds drop to the ground and normally germinate immediately.
2. Many tropical plants grown under conditions of high temperature and humidity; these include such plants as sugarcane, rubber, jackfruit, macadamia, avocado, loquat, citrus, many palms, litchi, mango, tea, choyote, cocoa, coffee, tung, and kola.
3. Many aquatic plants of the temperate zones, such as wild rice (*Zizania*), pondweeds, arrowheads, and rushes.
4. Many tree nut and similar species with large fleshy cotyledons, such as hickories and pecan (*Carya*), birch (*Betula*), hornbeam (*Carpinus*), hazel and filbert (*Corylus*), chestnut (*Castanea*), beech (*Fagus*), oak (*Quercus*), walnut (*Juglans*), and buckeye (*Aesculus*).

## Orthodox Seeds

The majority of important crop plants are species with **orthodox seeds.** Orthodox seeds tolerate drying after seed development and can be stored in a dry state (usually 4 percent to 10 percent moisture) for extended periods of time. Species with orthodox seed behavior vary in the length of time they tolerate storage.

Medium-Lived Seeds. Medium-lived seeds remain viable for periods of 2 or 3 up to perhaps 15 years, providing that seeds are stored at low humidity and, preferably, at low temperatures. Seeds of most conifers, fruit trees, and commercially grown vegetables, flowers, and grains fall into this group. Crop species can be grouped according to the ability of seeds to survive under favorable ambient storage conditions (Table –2). The Relative Storability Index indicates the storage time where 50 percent or more of seeds can be expected to germinate. Seed longevity will be *considerably longer* under controlled low temperature and humidity storage.

Long-Lived Seeds. Many of the longest-lived seeds have hard seed coats that are impermeable to water. Plant families that produce seeds with hard seed coats include the legume, geranium, and morning glory families. If the hard seed coat remains undamaged, such seeds can remain viable for at least 15 to 20 years. The maximum life can be as long as 75 to 100 years and perhaps more. Records exist of seeds being kept in museum cupboards for 150 to 200 years while still retaining viability. There are a number of claims of seeds from ancient tombs germinating after thousands of years. However, these lack definitive scientific support. Indian lotus (*Nelumbo nucifera*) seeds that had been buried in a Manchurian peat bog were originally estimated to be more than 1,000 years old and germinated perfectly when the impermeable seed coats were cracked. However, recent carbon-14 dating of these and other lotus seeds estimate the age of these seeds to be only 100 to 430 years old !

## *Storage Factors Affecting Seed Germination*

As seeds **deteriorate,** they:

1. first **lose vigor,**
2. then the **capacity for normal germination,**
3. and finally **viability.**

Storage conditions that reduce seed deterioration are those that slow respiration and other metabolic processes without injuring the embryo. The most important conditions are low moisture content of the seed, low storage temperature, and modification of the storage atmosphere. Of these, the moisture-temperature relationships have the most practical significance.

Harrington introduced a "rule of thumb" that indicated that seeds lose half their storage life for every 1 percent increase in seed moisture between 5 percent and 14 percent. Also, seeds lose half their storage life for every 5°C increase in storage temperature between 0 and 50°C. This is, of course, a generalized theory that varies between species. More accurate mathematical models have been developed to predict seed longevity at various temperature and moisture contents.

The most important factors impacting extended seed longevity in storage are seed moisture content and storage temperature.

**Moisture Content** Control of seed moisture content is probably the most important factor in seed longevity and storage. Most crop species have orthodox seeds where dehydration is their natural state at maturity. These seeds are best stored at a non-fluctuating low moisture content.

Seeds of orthodox species are desiccation-tolerant and, for most, 4 percent to 6 percent moisture content is favorable for prolonged storage , although a somewhat higher moisture level is allowable if the temperature is reduced. For example, for tomato seed stored at 4.5 to 10°C (40 to 50° F), the percent moisture content should be no more than 13 percent; if 21°C (70°F), 11 percent; and if 26.5°C (80°F), 9 percent.

Various storage problems arise with increasing seed moisture. At 8 percent or 9 percent or more, insects are active and reproduce; above 12 percent to 14 percent, fungi are active; above 18 percent to 20 percent, heating may occur due to seed respiration; and above 40 percent to 60 percent, germination occurs.

If the moisture content of the seed is too low (1 percent to 2 percent), loss in viability and reduced germination rate can occur in some kinds of seeds.

For seeds stored at these low moisture levels, it would be best to rehydrate with saturated water vapor to avoid injury to seed. Moisture in seeds is in equilibrium with the relative humidity of the air in storage containers, and increases if the relative humidity increases and decreases if it is reduced.

Thus, moisture percentage varies with the kind of storage reserves within the seed. Longevity of seed is best if stored at 20 percent to 25 percent relative humidity. Since fluctuations in seed moisture during storage reduce seed longevity , the ability to store seeds exposed to the open atmosphere varies greatly in different climatic areas. Dry climates are conducive to increased longevity; areas with high relative humidity result in shorter seed life. Seed viability is particularly difficult to maintain in open storage in tropical areas.

Storage in hermetically sealed, moisture-resistant containers is advantageous for long storage, but seed moisture content must be low at the time of sealing. Seed moisture content of 10 percent to 12 percent (in contrast to 4 percent to 6 percent) in a sealed container is worse than storage in an unsealed container.

Recalcitrant seeds owe their short life primarily to their sensitivity to low moisture content. For instance, in silver maple (*Acer saccharinum*), seed moisture content was 58 percent in the spring when fruits were released from the tree. Viability was lost when moisture content dropped below 30 percent to 34 percent.

Citrus seeds can withstand only slight drying without loss of viability. The same is true for seeds of some water plants, such as wild rice, which can be stored directly in water at low temperature. The large fleshy seeds of oaks (*Quercus*), hickories (*Carya*), and walnut (*Juglans*) lose viability if allowed to dry after ripening.

Viability of recalcitrant seeds of the temperate zone can be preserved for a period of time if kept in a moist environment and the temperature is lowere. Under these conditions many kinds of seeds can be kept for a year or more. Seeds of some tropical species (e.g., cacao, coffee), however, show chilling injury below 10°C (50°F).

**Temperature** Reduced temperature invariably lengthens the storage life of seeds and, in general, can offset the adverse effect of a high moisture content. Subfreezing temperatures, at least down to –18°C (0°F), will increase storage life of most kinds of seeds, but moisture content should not be high enough to allow the free water in the seeds to freeze and cause injury. Refrigerated storage should be combined with dehumidification or with sealing dried seeds in moisture-proof containers.

Cryopreservation. Survival of seeds exposed to ultralow temperatures (**cryopreservation**) has been known since 1879. There is renewed interest in storage of seeds by cryopreservation because it is potentially a cost-effective way to preserve germplasm for long periods of time with minimal loss of genetic information due to chromosomal mutations that accompany seed deterioration. Seeds are cryopreserved by immersion and storage in liquid nitrogen at –196°C (Fig. –34). Seed moisture must be low for survival, and gradual cooling and warming rates limit damage to the seed like cracks in the seed coat.

Cryopreservation of seeds has not replaced standard long-term storage at –18°C because long-term effects on seed survival have yet to be determined. However, numerous species have been stored for short periods of time in liquid nitrogen with promising results. Research is continuing, especially at the National Seed Storage Lab to make cryopreservation an important tool for seed preservation. Cryopreservation technology is also being applied to other tissue like pollen and dormant buds for possible preservation of germplasm.

## *Types of Seed Storage*

Although optimal seed storage conditions are cold temperature and low relative humidity, it is not always possible to maintain these conditions for commercial seed lots because of economic reasons. Typical conditions for commercial storage listed from least to most expensive include: (Fig. –35)

1. **Open storage** without humidity or temperature control
2. Storage in **sealed containers** with or without temperature control
3. **Conditioned storage** with humidity and temperature control

*Figure –34 Germplasm storage. (a) Movable storage cabinets for seed storage. (b and c) Seed storage in liquidnitrogen- filled dewers.*

*Figure –35 Various seed storage methods. (a) Small, high value seeds in plastic containers. (b) Vegetable seeds stored in sealed cans. (c) Large-seeded vegetables in bulk storage in waxed boxes. (d) Conditioned storage for crop seeds. (e) Refrigerator storage for flower seeds.*

## Open Storage without Humidity or Temperature

Control Many kinds of orthodox seeds need to be stored only from harvest until the next planting season. Under these conditions, seed longevity depends on the relative humidity and temperature of the storage atmosphere, the kind of seed, and its condition at the beginning of storage. Basic features of the storage structures include (a) protection from water, (b) avoidance of mixture with other seeds or exposure to herbicides, and (c) protection from rodents, insects, fungi, and fire.

Retention of viability varies with the climatic factors of the area in which storage occurs. Poorest conditions are found in warm, humid climates; best storage conditions occur in dry, cold regions. Fumigation or insecticidal treatments may be necessary to control insect infestations.

Open storage can be used for many kinds of commercial seeds for at least a year (i.e., to hold seeds from one season to the next). Seeds of many species, including most agricultural, vegetable, and flower seeds, will retain viability for longer periods up to 4 to 5 years , except under the most adverse conditions.

**Sealed Containers** Packaging dry seeds in hermetically sealed, moisture-proof containers is an important method of handling and/or merchandising seeds. Containers made of different materials vary in durability and strength, cost, protective capacity against rodents and insects, and ability to retain or transmit moisture. Those completely resistant to moisture transmission include tin or aluminum cans (if properly sealed), hermetically sealed glass jars, and aluminum pouches.

Those almost as good (80 percent to 90 percent effective) are polyethylene (3 mil or thicker) and various types of aluminum-laminated paper bags. Somewhat less desirable, in regard to moisture transmission, are asphalt and polyethylene-laminated paper bags and friction-top tin cans. Paper and cloth bags give no protection against moisture change. Small quantities of seeds can be stored satisfactorily in small moisture-proof containers like mason jars or plastic food containers.

Seed may be protected against moisture uptake by mixing with a desiccant. A useful desiccant is silica gel treated with cobalt chloride. Silica gel (one part to ten parts seed, by weight) can absorb water up to 40 percent of its weight. Cobalt chloride turns from blue to pink at 45 percent RH and can act as a useful indicator of excess moisture. Seeds should not be stored in contact with the desiccant. Seeds in sealed containers are more sensitive to excess moisture than when subjected to fluctuating moisture content in open storage. Seed moisture content of 5 percent to 8 percent or less is desirable, depending on the species.

**Conditioned Storage** Conditioned storage includes use of dehumidified and/or refrigerated facilities to reduce temperature and relative humidity. Such facilities are expensive but are justified where particularly valuable commercial seeds are stored. It is also justified for research, breeding stocks, and germplasm.

Also in some climatic areas, such as in the highly humid tropics, orthodox seeds cannot be maintained from one harvest season to the next planting season.

Cold storage of tree and shrub seed used in nursery production is generally advisable if the seeds are to be held for longer than 1 year. Seed storage is useful in forestry because of the uncertainty of good seed-crop years. Seeds of many species are best stored under cold, dry conditions. Ambient relative humidity in conditioned storage should not be higher than 65 percent to 75 percent RH (for fungus control) and no lower than 20 percent to 25 percent.

It is important to control humidity in refrigerated storage since the relative humidity increases with a decrease in temperature and moisture will condense on the seed. At 15°C (59°F), this equilibrium moisture may be too high for proper seed storage. Although the seed moisture content may not be harmful at those low temperatures, rapid deterioration will occur when the seeds are removed from storage and returned to ambient uncontrolled temperatures. Consequently, refrigeration should be combined with dehumidification or sealing in moisture-proof containers.

Low humidity in storage can be obtained by judicious ventilation, moisture proofing, and dehumidification as well as by the use of sealed moisture containers, or the use of desiccants, as described previously.

Dehumidifiers utilize desiccants (silica gel) or saturated salt solutions. The most effective storage is to dry seeds to 3 percent to 8 percent moisture, place in sealed containers, and store at temperatures of 1 to 5°C (41°F). Below-freezing temperatures can be even more effective if the value of the seed justifies the cost.

Moist, Cool Storage for Recalcitrant Seeds. Many recalcitrant seeds that cannot be dried can be mixed with a moisture-retaining medium, placed in a polyethylene bag or other container, and refrigerated at 0 to 10°C (32 to 50°F). The relative humidity in storage should be 80 percent to 90 percent. Examples of species whose seeds require this storage treatment are silver maple (*Acer saccharinum*), buckeye (*Aesculus* spp.), American hornbeam (*Carpinus caroliniana*), hickory (*Carya* spp.), chestnut (*Castanea* spp.), filbert (*Corylus* spp.), citrus (*Citrus* spp.), loquat (*Eriobotrya japonica*), beech (*Fagus* spp.), walnut (*Juglans* spp.), litchi, tupelo (*Nyssa silvatica*), avocado (*Persea* spp.), pawpaw (*Asimina triloba*), and oak (*Quercus* spp.). The procedure is similar to moist-chilling (stratification).

Acorns and large nuts may be dipped in paraffin or sprayed with latex paint before storage to preserve their moisture content.

# Chapter 2

# TROPICAL AND SUBTROPICAL FRUITS

## CITRUS FRUITS (*CITRUS* SPP.)

### *Introduction*

The major cultivated types of citrus fruits included in the genus *Citrus* are broadly grouped into oranges, mandarin, lemons, limes, grapefruit, and pummelo. The genus is cultivated for sweet and acidic fruits. Sweet orange, mandarin, and grapefruit are eaten fresh or processed for squash (sweetened citrus fruit juice) and juice preparation. Lemon and lime are acidic in nature and largely used in preparation of culinary products, such as pickles, and for flavoring food items. They are also processed for juice, squash, and lemonade.

Citrus fruits are a rich source of sugar, citric acid, and vitamin C, and they possess valuable medicinal properties, being used in the prevention of colds and malaria, and to promote blood coagulation. The wide genetic diversity for citrus fruits exists in Southeast Asia, and they are predominantly distributed in the Indo-Chinese region. They grow well in tropical and subtropical regions. Some major citrus growing countries are Argentina, Australia, Brazil, China, Egypt, Greece, India, Israel, Italy, Japan, Mexico, Morocco, South Africa, Spain, Turkey, the United States, and the West Indies.

### *Morphology*

Citrus fruits belong to family Rutaceae, subfamily Aurantioideae, and tribe Citreae. The tree is an evergreen, dicotyledonous perennial with spines. Leaves are simple ovate with wings at the petiole region. Flowers are single or in clusters, white, and fragrant. Fruit is a hesperidium, carpels, or segments filled with juicy arils and seeds. Seeds are white, show poly embryony, excepting *Citrus grandis,* and vary in size and number in different species.

## Seed Storage

Citrus plants are propagated both sexually and asexually. Seeds of different *Citrus* species exhibit polyembryony, where nucellar seedlings are vigorous, true to type, and commonly used for raising citrus plants. In vegetative propagation, a large number of rootstocks is used to impart better fruit quality and resistance to various biotic and abiotic stresses in scionic plants. These are selected based on better compatibility with scion, hardy nature, and tolerance to saline conditions or soilborne diseases. They are raised exclusively through seeds. Seeds also play an important role in evolving new cultivars through hybridization, and in long-term conservation of genetic diversity.

High seed viability can be preserved for a shorter period for plant propagation, and for a longer period for the conservation of genetic diversity in the gene bank. Earlier, seeds of different *Citrus* species showed various storage behaviors, recalcitrance to orthodox under different storage conditions.

However, King and Roberts grouped them as recalcitrant, wherein drying and freezing kill seeds. Drying of citrus seeds caused considerable delay in radicle emergence, thereby requiring a longer period for full seedling emergence. Further, rapid seed drying caused desiccation injury at higher temperatures. Earlier seeds were germinated for a short period, and delayed seedling emergence due to seed drying was mistaken for loss of viability. In many *Citrus* species, seed germination is slow and erratic, owing to both physical and chemical inhibiting factors, and the undifferentiated embryo is likely to be a cause for delayed germination. It is still unclear in certain *Citrus* species whether they belong to the orthodox or recalcitrant category. Hong and Ellis opined that interspecific variation for seed storage behavior exists within the genus *Citrus,* and the prevalence of an intermediate category of storage behavior is not ruled out between orthodox and recalcitrant types, similar to one that occurs in *Coffea* spp.

### *Seed Collections*

Citrus is predominantly cultivated in subtropical regions, between 40° north and south of the equator. Young trees are susceptible to frost and grow well in sandy loam soil. Plants are raised almost exclusively by budding on desirable rootstocks and seldom through nucellar seedlings. The latter ones are more vigorous than sexual seedlings. *Citrus* species are cross-pollinated, and pollination is mainly by insects. Fruits develop slowly and take longer to mature. Seeds are extracted by cutting the fruits, then washed thoroughly, dried in shade or at low temperature, and suitably packed for storage. Washing seeds immediately after their extraction from fruits generally gives better germination than leaving seeds in rotting fruits. It was reported by Elze that the storage of picked fruits for several weeks affected germination.

## SOUR ORANGE: CITRUS AURANTIUM L.

Sour orange, also called Seville orange, is commonly used in preparation of marmalade. Fruit is bitter and has medicinal properties. Sour orange is widely used as rootstock for sweet orange, lemon, and grapefruit due to its resistance to gummosis. It is a native of Southeast Asia and has somatic chromosome number 2n = 18. Tree is of medium size, growing to 10 meters (m) in height. Leaves are ovate, medium sized, and broadly winged. Flowers are large, axillary, white, and fragrant; stamens number 20 to 25; and the ovary has 10 to 12 locules. Fruit is subglobose, with a peel that is thick and rough, aromatic, and bright orange at maturity; the pulp is sour and bitter and embedded with many seeds. Seeds are polyembryonic and exhibit a high proportion of nucellar embryos.

### Seed Storage

#### *Seed Germination*

Sour orange seeds germinate at a constant temperature ranging from 15 to 38°C. Seedling emergence is rapid and higher at 30 to 35°C. According to Monselise, optimum temperature for seed germination is 26°C.

Seed viability is higher immediately after removal from ripe fruits. Seed shows progressive reduction in percentage of germination on air-drying or on storage. Burger noted that drying seed even for one day delays and decreases germination. Seed drying slowly at the rate of 30 percent moisture loss per 100 h imparted best viability. Seeds selected from hardy trees are more vigorous and show higher percentage of polyembryony.

Seed germination is relatively poor in sour orange and is attributed to the presence of inhibiting substances. Removal of testa and tegmen resulted in higher uptake of oxygen and gave higher percentage of germination.

Furthermore, seed germination improves with seed soaking in water, in gibberellic acid at 50 parts per million (ppm) and at 200 ppm , and in magnesium sulfate or magnesium citrate (1 percent) ; with cold stratification (5°C for 45 days) ; and upon priming in polyethylene glycol (– 0.90 megapascals [MPa] for 30 min).

### *Storage Conditions*

The viability of sour orange seeds decreases rapidly under ambient storage conditions , with seeds remaining viable for 120 days with or without seed coat removal. Seed viability declines further on exposure of fruit to chilling temperatures (– 7.8°C for 12 h). It was earlier believed that seed loses viability rapidly on drying, and, therefore, storage in fruit juice, phenols, growth regulators, fungicides, and solutions of high osmotic potential would be also ineffective for preserving viability. However, seed stored at 4°C prevented germination and growth of microorganisms, and survival of imbibed seeds was best in 10 – 2 M (molar concentration) indoleacetic acid at 4°C. Edwards and Mumford observed the longest survival of sour orange seeds in air at 4°C. Seed requires relatively small amount of oxygen to maintain viability during dry storage of seeds, and removal of moisture is beneficial in maintaining high viability for longer periods. King, Soetisna, and Roberts concluded that seed remains viable for eight months when dry seeds (5 percent mc) are stored at – 20°C.

## SWEET ORANGE: CITRUS SINENSIS (L.) OSBECK

With its origin in Southern China sweet orange is a major fruit crop cultivated extensively in the citrus industry. The diploid chromosome number is 2n = 18. Fruit is eaten fresh as a dessert or snack. Tree is of medium size and grows up to 12 m high. Leaves are medium size with narrowly winged petioles. Flowers are axillary, single or in clusters, fragrant and white; stamens number 20 to 25; and the ovary has 10 to 14 locules. Fruit is subglobose and oval, with a tightly held peel, which turns orange on ripening, though it remains green in the tropics. Seeds are white, nil to many, and polyembryonic.

## Seed Storage

### *Seed Germination*

Sweet orange is propagated by vegetative means and requires rootstocks with straight roots and shoots for ideal grafting. Sweet orange seeds germi nate at a temperature range of 15 to 37°C. Seed germination is slow and often erratic. No germination of seeds occurs for the first 15 days of sowing, but subsequently it continues up to 31 days at 23°C. Seed germination is better at an alternate temperature of 20/30°C for 16/8 h. Benching affects the uptake of nutrients and translocation of food reserve to fruits. The seed coat contains inhibitors, which was confirmed by inhibition of cress and lettuce seed germination. Removal of the seed coat prevents benching of the stem , bending of roots , and gives higher percentage of germination.

### Storage Conditions

Higher seed moisture reduces the storage life. Well-dried seeds packed in polyethylene bags stored at 5.5°C maintain high viability (72 percent) for 12 months. Seeds did not germinate when moisture content was less than 6 percent. The drier the seeds, the greater the delay in germination. Bajpai, Trivedi, and Prasad opined that leaving seeds in whole fruits is better than storage in polyethylene bags, and seed storage in charcoal is best for sweet orange. Seeds remained in good condition for several months when packed in sealed polyethylene bags. According to Horanic and Gardener , low temperature affects seed viability during storage. Seeds stored at – 3.9°C for 24 h were not injured but were partially damaged at – 6.7°C and killed at – 9.4°C.

# MANDARIN: CITRUS RETICULATA BLANCO.

Mandarin is commonly referred to as loose-skinned orange. Fruit is sweet and subacidic and eaten fresh as a dessert or snack. It is also used for preparation of juice, squash, syrup, and lemonade. Mandarin originated in China, but its cultivation has extended to other tropical and subtropical regions. Mandarin is the hardiest among cultivated *Citrus* species. Trees are small to medium size, with a few thorns; leaves are small and have narrowly winged petioles. Flowers are small and white, stamens number 20, and the ovary has 10 to 15 locules. Fruit is depressed globose, with a thin peel; has easily separable segments; is orange-red, green, or yellow on ripening; and has sweet, juicy orange pulp. Seeds are small, pointed, and polyembryonic with green cotyledons. The somatic chromosome number is 2n = 18.

## Seed Storage

### Seed Germination

Mandarin seeds germinate at 15 to 35°C , and a constant temperature of 25°C is ideal for germination. Seed germination is affected by saline conditions, decreasing with increase in soil salinity. Seedling emergence is rapid and higher in seeds treated with ascorbic acid (100 ppm).

### Storage Conditions

Mandarin seeds survive for a shorter period under ambient conditions. High seed moisture at higher storage temperature promotes metabolism and encourages microbial growth in seeds, leading to loss of viability. Seeds of *Citrus* species behave differently toward desiccation and dry storage. However, they are dried slowly and carefully as they attain a more advanced stage of germination; otherwise, they are killed on desiccation at a higher level of moisture. Seed germination is reduced from 95 to 65 percent after two days of storage under room temperature, and seed moisture is concurrently reduced from 50.9 to 12.2 percent. Rapid reduction of germination might be due to desiccation injury.

Loss of seed viability in *Citrus reticulata* is accompanied by a decline in nucleic acid levels; protein hydrolysis; protease, catalase, and peroxidase activity; and increases in leaching of metabolites. The quantum of leachate increases with an increase in storage period.

Mungomery, Agnew, and Prodonoff suggested preserving mandarin seeds with initial moisture in a cool place and protecting them with suitable fungicides for longer storage. The percentage of germination was 24 in seeds stored at 4.5°C after 80 days of storage. Seeds stored in fruit also retain the viability for 120 days at 15°C without affecting seed and fruit quality.

## LEMON: CITRUS LIMON BURM.

Lemons are largely used in preparation of squashes, juices, lemonades, pickles, and various confectionery items. Lemon oil, citric acid, and pectin are extracted from the fruit and used in cosmetics. Lemon is native to Northeastern India. It is a small tree of about 3 to 6 m height with stiff thorns. The somatic chromosome number is 2n = 18. Leaves are simple and ovate and petioles are not winged. Flowers are single or in clusters, axillary, pink white; stamens number 20 to 40; and the ovary has eight to ten locules. Fruits are light yellow, oval, with a thick peel and terminal nipple, and sour. Seeds are ovoid and polyembryonic with white cotyledons.

### Seed Storage

#### *Seed Germination*

Seed germination is slow and uneven in lemons due to their hard seed coat. The extract of lemon seed coat inhibits germination of cress and let- tuce seeds. Seed germination is further delayed and affected by drying, which requires longer imbibitions during germination. The optimum temperature for imbibitions is 35 to 40°C. At higher storage temperature, deterioration is rapid when moisture is more than 5 percent, and more abnormal seedlings are formed. The rate of seed germination increases with removal of the hard seed coat. King and Roberts reported that high seed moisture is necessary to maintain high viability during storage, and 22 percent mc or less reduces the percentage of germination. However, Mumford and Grout reported that *Citrus limon* seeds were tolerant to extreme desiccation (1.2 percent mc) and that dry seeds showed no changes in germination percentage even after freezing.

### Storage Conditions

King, Soetisna, and Roberts confirmed that lemon seeds are orthodox in storage behavior and can be dried to a low level of moisture without affecting the viability. Furthermore, seeds with 5 percent moisture maintained viability for seven months at – 20°C. Lemon seed viability is preserved well with lowering of seed moisture and under cool conditions. High seed viability is preserved for 40 days at room temperature and for 60 days at 5°C when seed is stored in the fruit itself. Seeds without testa stored at –18°C retained their viability for 655 days.

## ROUGH LEMON: CITRUS JAMBHIRI LUSH.

Rough lemon is indigenous to India and predominantly used as rootstock for cultivated *Citrus* species. The somatic chromosome number is 2n = 18. Trees are medium size with few thorns. Leaves are dark green and pointed; flowers are small; and fruit is rough and round, has an irregular surface, and contains many small seeds.

## Seed Storage

### Seed Germination

Seed viability is comparatively lower in rough lemon than in other rootstocks. Fresh seeds of Florida rough lemon cultivar gave only 20 percent germination. Ali noted higher germination in descending order of rootstocks of rough lemon, smooth lemon, *Citrus karna, C. aurantifolia,* and *C. aurantium.* Seed germination can be improved by soaking seed in gibberellic acid (500 ppm) , and through cold stratification of seeds for 45 days at 5°C.

### Storage Conditions

Seed viability decreases with an increase in storage period. Seed germination reduced to 31 percent after four months of storage. Seed viability can be preserved for a longer period under cool and dry storage conditions. Seeds stored at 4°C and 96 percent RH exhibited 45 percent germination after 12 months of storage.

### Invigoration of Stored Seeds

Seed vigor decreases during improper seed storage. Seed soaking in gibberellic acid (10 ppm) enhances germination from 31 to 76 percent even after four months of storage.

## LIME: CITRUS AURANTIFOLIA (CHRISTM.) SWINGLE

Lime is widely cultivated in Asia for its edible fruits. Fruits are used for pickles, and their juice is used for flavoring food items and in the preparation of syrup and concentrate. Lime oil and citric acid are extracted from the fruit. Lime originated in northern parts of India and has somatic chromosome number 2n = 18. Lime trees are small with sharp spines, reaching up to 5 m in height. Leaves are small, ovate, with narrowly winged petioles.

Flowers are small and white, stamens number 20 to 25, and the ovary has 9 to 12 locules. Fruits are small, round to oval, with a thin peel, aromatic, greenish yellow, and highly acidic. Seeds are white and polyembryonic.

## Seed Storage

### *Seed Germination*

Seed germination is higher at a constant temperature of 25°C. Large and extralarge fruits give bold seeds with higher germination and vigorous seedlings. The seed coat contains inhibitors that prevent germination, and its removal hastens the germination process. Seed germination improves with soaking seeds in gibberellic acid (40 ppm), naphthalene acetic acid (40 ppm), thiourea (1.5 to 2.0 percent), and potassium nitrate (2 percent).

## Storage Conditions

Seed longevity varies under different storage conditions. Mumford and Panggabean stored seeds for more than 120 days at room tempera ture. However, Doijode reported that germination declines rapidly on drying. It was 60 percent after 60 days of storage and lost completely by 365 days of ambient storage.

High seed viability and vigor are preserved for 120 days when seeds are stored in fruit at 5 or 15°C . Bajpai, Trivedi, and Prasad reported that lime seeds could be stored best in charcoal powder at 3°C and 50 percent RH. Lime seeds with 5 percent moisture stored at –20°C preserve viability for eight weeks. When storage is extended to 270 days at 4°C, viability is 90 percent. Low moisture seeds (6.7 percent) stored in polyethylene and laminated aluminum foil pouches at 5 and –20°C preserve seed viability for three years. The retention of viability is greater in laminated aluminum foil pouches than in polyethylene bags.

Seed longevity can also be extended further with low-temperature (– 20°C) storage, with 43 percent germination recorded after 12 years of storage. Seed viability preserves well under modified atmosphere storage, especially in nitrogen.

# RANGPUR LIME: CITRUS LIMONIA OSBECK

Rangpur lime originated in India and is mainly used as a rootstock for citrus fruits in India and Brazil. The diploid chromosome number is 2n = 18. It is a hardy and drought-tolerant rootstock. Trees are medium sized with few thorns, with dull-green leaves and flowers that are tinged with purple. Fruits are round to depressed globose with a short nipple, yellow to reddish orange, with a moderately loose peel; they have eight to ten segments, give acidic juice, and contain numerous seeds. Large fruits contain a greater number of heavy seeds, while unripe fruits contain few bold seeds.

## Seed Storage

### *Seed Germination*

Seed germination is higher at a constant temperature range of 21 to 38°C. Seed germination decreases with fruit size. Seeds are placed pointing downward or sideways to get early and more straight seedlings. Seed viability and vigor reduce on drying. The degree of reduction varies in different genotypes. It was higher in Brazilian rangpur lime, followed by the Florida strain. It appears that initial loss of viability is due to desiccation injury.

### *Storage Conditions*

Rangpur lime seeds were stored better at 10°C and 45 percent RH for six weeks. Seed longevity can be extended further to 14 months when seeds are stored in sealed containers at 2 to 3°C and 33 to 38 percent RH.

# GRAPEFRUIT: CITRUS PARADISI MACFAD.

Grapefruit is largely cultivated for the edible juicy fruits. Fruits possess a characteristic flavor with mild bitterness and are largely used for juice. It probably originated in the West Indies. The somatic chromosome number is 2n = 18. Trees are large and spiny with dense foliage of large, dark green leaves with winged petioles. Flowers are white, large, single or in clusters; stamens number 20 to 25; the ovary has 12 to 14 locules. Fruit is large, globose, and yellow and contains many seeds. Seeds are white and polyembryonic.

## Seed Storage

### *Seed Germination*

Seed germination is higher at a constant temperature range of 15 to 38°C and at alternate temperatures of 20/30°C (16/8 h). It improves with removal of seed coat. Seeds extracted from immature fruit do not germinate. The germination is slow and delayed on drying , and it improves with soaking seeds in potassium dichromate at 0.5 percent.

### Storage Conditions

Seed moisture plays an important role in deciding storage life of seeds. Low-moisture seeds stored in sealed conditions maintained viability for 16 weeks at room temperatures. Seeds stored in open conditions at room temperatures and at 5 and 8°C retained viability for three and four months, respectively. Seed viability over 50 percent can be retained for six months when seeds are stored at 5 and 8°C in sealed polyethylene and mixed with charcoal, ash, sand, or sawdust. Seed germination was 40 percent when stored at 4.5°C in sealed plastic bags after 80 days of storage. Seed viability was reduced on drying, whereas moist storage gave 80 percent germination. Well-dried seeds packed in polyethylene bags stored at 5.5°C showed 82 percent germination after 28 months of storage.

# PUMMELO: CITRUS GRANDIS OSBECK

Pummelo is cultivated for edible fruits largely in East Asia. Fruits are eaten fresh and used in the preparation of mixed fruit drinks and marmalade. It is native to Malaysia and Thailand. Trees are medium in size and about 5 to 15 m in height. Leaves are large, ovate, and dark green. Flowers are large, cream colored, single or in clusters, and fragrant; stamens number 20 to 25; the ovary has 11 to 16 locules. Fruit is large, pear shaped or globose, and yellow; it has a thick peel and sweet juice; and it contains few seeds. Seeds are large, yellow, and monoembryonic. The diploid chromosome number is $2n = 18$.

## Seed Storage

### Seed Germination

Pummelo seeds germinate well if they contain high moisture (15 to 20 percent). The reduction of seed moisture to 10 percent or less causes loss of viability. A minimum of 20 percent seed moisture is required to maintain high viability during storage of pummel seeds.

### Storage Conditions

Pummelo seeds need special attention during storage, being somewhat sensitive to drying. Also, a higher storage temperature reduces their viability. Seeds stored in sealed polyethylene bags at 4.4°C and 56 to 58 percent RH preserved their viability for 80 days.

# TRIFOLIATE ORANGE: PONCIRUS TRIFOLIATA (L.) RAF.

Trifoliate orange is most commonly used as a hardy rootstock for citrus. It is cold hardy, induces dwarfing, and is fairly resistant to soilborne diseases and nematodes.

It adapts to heavy soil and improves the scion fruit quality. Trifoliate orange fruits are nonedible, bitter, and used in medicine.

Different plant parts are showy, and the plant is thus used in home gardens as an ornamental. It probably originated in Central and Northern China. It grows best in warm climates and in heavy soil. It does not grow well under saline conditions. Trees are small with thorns and three-lobed leaves; flowers are small, sessile, and white; stamens number 20 or more; the ovary has six to eight locules. Fruit is small, with a rough surface of lemon color and filled with acidic juice, and it contains numerous seeds. Seeds are ovoid, plump, and polyembryonic.

## Seed Storage

### *Seed Germination*

Trifoliate orange seeds germinate at various temperatures, ranging from 12 to 35°C. Seed germination is reduced by 20 to 25 percent at lower temperatures. Minimum and optimum temperatures are 6 to 10°C and 30°C, respectively. Seed viability is affected by drying, and it decreases when moisture is reduced to less than 20 percent , 22 percent , and 27 percent. The percentage of germination increases with removal of the seed coat , washing in running water for 24 h , and soaking in gibberellic acid (10 ppm). Seedling emergence is slow and delayed in fresh seeds, exhibiting 24 percent germination.

### *Seed Maturity*

Seed maturity coincides with fruit maturity. The percentage of germination is rather low before maturity. Trifoliate orange seeds collected 90 to 105 days after anthesis germinate. The germination is maximum at 135 days after flowering. Embryo size plays a major role in seed germination. Reduction in embryo size lowers the percentage of germination. Germination of seeds extracted from yellow ripe fruits is comparatively lower than that for seeds of overripe fruits. However, seeds extracted at different stages of fruit development require storage for 12 weeks at 4.5°C for higher viability.

### *Storage Conditions*

Trifoliate orange seeds are short-lived, and their viability decreases with loss of moisture. A minimum of 25 percent moisture for storage of trifoliate seeds is recommended. Fresh seeds show 80 percent germination, and dry seeds stored at 4°C exhibited 3 percent germination, while moist seeds under similar conditions exhibited 50 percent germination after one year of storage. Seeds with 27 percent moisture stored in plastic bags at 8°C retained viability for 224 days. Seeds stored in a mixture of wet sand and vermiculite (50:50) at 5°C remained viable for 240 days but failed to germinate when stored in paper bags at ambient temperatures.

Seeds stored at low temperature (4°C) and high humidity (96 percent RH) exhibited 32 percent germination after 12 months of storage. The phenolic substances increased in seeds during storage. Okazaki, Ogawara, and Ino reported that seed longevity increased with open or sealed storage at low temperature or with storage in sand. Seeds stored in damp sand and at higher temperatures deteriorated rapidly. A well-dried seed material stored in polyethylene bags at 5.5°C gave 91 percent germination after 20 months of storage.

# Chapter 3

# MANGO: *MANGIFERA INDICA L.*

## INTRODUCTION

Mango is an important tropical fruit largely cultivated in Southeast Asia for its delicious fruit. It is most popular in India. Ripe fruits are eaten fresh and used in various products, such as juice, squash, jams, and jellies, and products such as chutney, pickles, and curries are made from unripe fruits. Fruits are rich in vitamins Aand C, and kernels contain high amounts of carbohydrates, calcium, and fat.

### Origin and Distribution

Mango originated in the Indo-Burma region. *Mangifera* species are distributed throughout the Indo-Malayan region, showing wide genetic diversity for different plant characters. Subsequently it was introduced to other tropical countries. It is cultivated in Australia, Bangladesh, Brazil, China, Colombia, Cuba, Egypt, Fiji, India, Indonesia, Jamaica, Madagascar, Malawi, Malaysia, Mauritius, Myanmar, Nigeria, Peru, the Philippines, Portugal, Seychelles, South Africa, Sri Lanka, Sudan, Thailand, the United States, Venezuela, and the West Indies.

### Morphology

Mango belongs to family Anacardiaceae and has chromosome number 2n = 40. It is a medium-sized, evergreen, dicotyledonous, perennial tree. Leaves are simple and dark green. Inflorescence is large and terminal; panicle bears about 300 to 1,000 staminate and perfect flowers. Fruit is an indehiscent, fleshy drupe and varies in size, shape, and color, being either green, yellow, or red. Pulp is soft, flavored, juicy, and green, yellow, or orange in color. Seeds are enclosed in a hard endocarp, with two fleshy cotyledons and little or no endosperm. Most of the cultivars are monoembryonic, while a few show polyembryony, with 2 to 12 embryos, where both apomictic and zygotic embryos exist.

## SEED STORAGE

Mango is commercially propagated by asexual methods. However, seeds are used for raising uniform rootstocks and orchards with polyembryonic cultivars, ensuring true-to-type plants. Seedling trees are long-lived and deep-rooted and show large variations in bearing habit and fruit characteristics. They are relatively late bearers and difficult to manage in large numbers. Normally, seeds are used for raising rootstocks and in hybridization programs. Mango seeds remain viable for short periods and exhibit recalcitrant storage behavior. Seeds are susceptible to desiccation and are killed when moisture goes below a critical level. Further, they do not withstand chilling temperature during storage. Seeds stored under moist conditions remain viable only for a short period, from a few weeks to a few months.

### Seed Collections

Mango grows well in tropical conditions. Young and flowering trees are damaged by frost. It can grow up to 1,000 m altitude in areas but performs best up to 500 m altitude. It thrives well in deep sandy loam soils. It is commonly propagated asexually through graftage. In polyembryonic cultivars, seeds are used for raising plants as well as rootstocks. Mango is cross-pollinated by insects. Fruits are collected from healthy trees having the desired qualities (see Figure.1). Fully ripe and healthy fruits are selected, and the pulp is removed from the stones, washed thoroughly, and stored.

### Seed Germination

Mango seeds are nondormant and germinate readily under favorable conditions. They are enclosed by a hard endocarp, which inhibits or delays seedling emergence. Removal of seeds from stones increases germination. According to Sinnadurai , decorticating helps in early emergence of seedlings and is also beneficial in development of the nucellar seedlings in polyembryonic cultivars. Seeds of monoembryonic cultivars germinate readily and emerge earlier than those of polyembryonic cultivars. Seeds extracted from soft-pulp fruits give higher germination (76 percent) than those from firm fruits (54 percent), and seeds from firm fruits remain dormant for some time after planting.

Corbineau, Kante, and Come reported that germination occurs between 5 and 40°C, but the optimum temperature is 25 to 30°C. Seedlings grow well at 30°C, and higher temperature (40°C) is injurious to growing seedlings, while temperature below 15°C causes chilling injury. Giri and Chaudhary noted that seed germination and seedling vigor are positively related to seed weight. Large embryos give rapid emergence of vigorous seedlings.

*Figure 1. Mango Tree with Many Fruits*

## Storage Conditions

Mango seeds lose viability if dried immediately after extraction. Due to their recalcitrant nature, presently they can only be preserved for a short duration and under specific conditions. A critical level of moisture determines the longevity of seeds, and this varies in cultivars, stages of maturity, and growing conditions. Doijode reported that reduction of seed moisture below 25 percent in 'Alphonso' and 32 percent in 'Totapuri' lowered the percentage of germination and seedling vigor. Fresh seeds remain viable for four to five weeks under ambient conditions, while Ito and Atubra reported that seeds remained viable for ten days under dry storage. Loss of viability is greater in highmoisture seeds, and also at higher storage temperatures. Parisot opined that loss of water from stored seeds and loss of germination capacity were rapid at high storage temperature (20 to 30°C). Retention of high moisture during storage promotes fungal growth, which causes discoloration and loss of viability. High moisture also stimulates the sprouting of seeds during ambient storage. With these constraints in view, suitable storage conditions need to be created, but so far, not much has been achieved in this regard.

*Seed storage in fruit.* Mango seeds deteriorate rapidly on separation from the fruit. As long as seeds remain inside the fruit, they are germinable. How-ever, subsequent shriveling and rotting of fruits affect the seed quality during ambient storage. Storing fruits at low temperature causes chilling injury.

According to Simao , the best method is to store whole fruit at room temperature (22 to 25°C), allowing seeds to remain viable for 70 days. Storage at 5°C harms the fruit. Large-fruit seeds stored better than small-fruit seeds.

*Seed storage with charcoal.* Charcoal powder is a better storage medium for mango seeds than polyethylene bags, desiccators, refrigeration, or open jars. It maintains suitable moisture and protects the seeds from pathogens. With this, viability of 50 to 60 percent is retained up to 90 days. Bajpai and Trivedi  reported that seeds stored with charcoal in a desiccator at 50 percent RH gave 80 percent germination even after 70 days of storage. Likewise, Chacko and Singh  obtained 37.5 percent germination after 100 days of storage of seeds with charcoal at 20 to 25°C. However, viability was reduced to 17.5 percent after 120 days of storage. Similarly, about 40 percent of seeds germinated after 90 days of storage when stones were stored in polyethylene bags along with charcoal at 25°C. While in open conditions, only 12 percent of seeds germinated by the sixtieth day of storage.

*Seed storage in moist conditions.* Seeds stored under moist conditions at relatively lower temperatures preserve viability. Seeds of cultivar Dashehari, when stored in thin polyethylene bags (0.025 millimeters [mm]) after treatment with 1 percent 8-hydroxyquinolene sulfate at 15°C, showed 50 percent germination after 12 months of storage (see Figure 2). Seeds stored at 8°C did not germinate , and chilling injury occurs at very low temperatures.

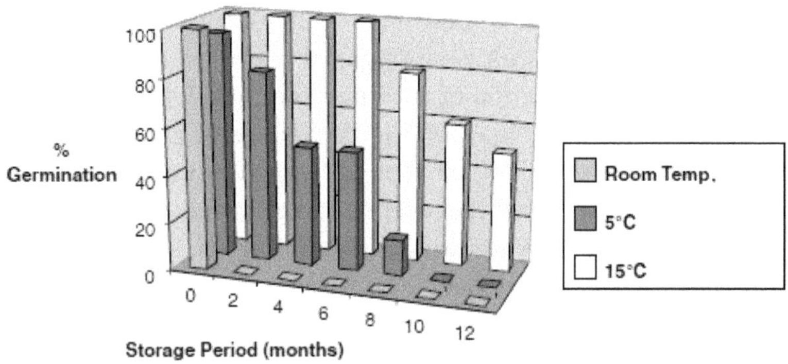


*Figure 2. Moist Storage of Mango Seeds*

The loss of seed viability is associated with structural changes in cells during deterioration, causing greater leakage of soluble sugars and amino acids and a decrease in dehydrogenase activity. Furthermore, Chandra reported a decline in RNA and reducing sugar in dry seeds with aging. Mango seeds having moisture above 20 percent maintain high viability at 15°C when packed in thin polyethylene bags and suitably protected against storage fungi.

# Chapter 4

# PAPAYA: *CARICA PAPAYA* L.

## INTRODUCTION

Papaya, also called pawpaw or melon tree, is a major dessert fruit crop in the tropics. Rich in carbohydrates, carotene, riboflavin, and vitamin C content, the fruit's sweet-flavored pulp is eaten fresh. It is also processed for jam, jelly, candy, and juice preparations. Different plant parts, particularly of the unripe fruit, yield a milky latex that contain the proteolytic enzyme papain, which has many industrial and medicinal uses. Unripe fruits are also used for pickle preparation and cooked as vegetables.

### Origin and Distribution

Papaya originated in Central America. Many of its wild species are grown in Colombia, Ecuador, Mexico, and Nicaragua. The papaya is cultivated throughout the tropical and subtropical regions. Some major papaya- growing countries are Australia, Brazil, Colombia, Costa Rica, Cuba, Ecuador, India, Kenya, Malaysia, Mexico, Myanmar, Nigeria, Peru, the Philippines, Seychelles, South Africa, Sri Lanka, Taiwan, the United States, and Venezuela.

### Morphology

Papaya belongs to family Caricaceae; its somatic chromosome number is 2n = 18. Papaya is a small herbaceous and dicotyledonous plant with a stem that is soft wooded, hollow, and straight; rarely branches; and ends with a cluster of foliage. Leaves are large lobed with long petioles. Plants show a pistillate habit that is stable, while staminate and andromonoecious trees produce staminate, pistillate, and hermaphrodite flowers, depending on climatic conditions. Storey  reported 31 sex types in papaya, with 15 each for variation in staminate and andromonoecious trees, and the final one pistillate. Staminate and andromonoecious trees are heterozygote, while the pistillate is a recessive homozygote. Fruit is a berry that is ovoid and oblong to spherical or cylindrical in shape and can be smooth or grooved. Flesh is yellow to reddish orange. Fruits vary in size and shape depending upon sex type and cultivars. Fruit cavity is filled with a mass of black, round seeds about 5 mm in diameter, surrounded by gelatinous sarcotesta; the embryo is straight with ovoid flattened cotyledons enclosed by fleshy endosperm. Seeds constitute about 16 percent of total fresh weight of fruit.

## SEED STORAGE

In papaya, seeds are predominantly used for raising plants. They are also used for evolving new cultivars through hybridization, and for long-term conservation of genetic diversity. Papaya seeds survive for a short period under ambient conditions. Seeds show orthodox storage behavior , in which seeds withstand desiccation and extend longevity at lower temperatures. Papaya seeds, however, are susceptible to chilling temperatures and killed when stored at zero or subzero temperatures. Therefore, seeds may be considered intermediate to recalcitrant with orthodox storage behavior.

### Seed Collections

Papaya is a warm-season crop. It requires hot sunny days for production of high-quality fruits. It is susceptible to frost, and low temperatures affect growth. Papaya grows well in deep, fertile, well-drained clay loamy soil that is free from nematodes. Either seeds are sown directly in the field or seedlings are raised in a nursery and subsequently transplanted in field. Seedling emergence is faster and higher when the plot is covered with cheesecloth. Flowering commences four to eight months after planting, and insects cause pollination. To maintain genetic purity of strains, flowers are hand pollinated and covered with bags to prevent pollen contamination. Fruit takes about five months to mature. It turns green to yellow on maturity (see Figure. 1). A greater fruit weight produces more seeds in fruits ripened during spring and autumn than in those ripened during summer. The seed mass is soaked in water for 24 h, and later seeds are separated from the pulp. Seeds are cleaned; mucilaginous material is removed, dried, and suitably packed for storage.

*Figure.1. Papaya Fruits Ready for Harvest*

*Fruit maturity.* Seed germination and longevity are dependent on seed maturity, which coincides with maturity of the fruit. Seed matures when the fruit becomes yellowish. Seed germination was highest (80 percent) in seeds collected from the least mature fruits (one-quarter mature) and stored for 33 days at room temperature.

*Seed drying.* Rapid drying of seeds affects germination, whereas sun drying does not affect seed viability. Longer duration of sun drying (>3 days) reduces germination. Seedling emergence is earlier and higher in seeds dried in shade compared to those dried in sun.

## Seed Germination

Fresh papaya seeds do not germinate readily. They exhibit dormancy up to a period of 35 days. In fresh seeds, seedling emergence is slow and nonuniform and only 6 percent of seeds germinate. Low seed germination is also partly due to the absence of an embryo in about 20 percent of seeds , and seed viability improves with a short period of storage. Large seeds give a slightly higher germination percentage than small seeds. Papaya seed germination is influenced by the following factors.

*Sarcotesta.* This is a gelatinous mucilaginous substance derived from the aril that surrounds papaya seeds. It protects the seeds from extraneous environment and delays germination. Seed germination is slower and lower in seeds with the sarcotesta intact than in seeds from which it is removed. Fresh seeds with the aril intact did not germinate for 30 days after sowing. Removal of the aril from seeds enhances germination to 40 to 50 percent, and further air-drying increases germination to above 80 percent. Reyes, Perez, and Cuevas reported the presence of inhibitors in sarcotesta and sclerotesta that affected germination. Seeds soaked in seed leachate or extracts of sarcotesta or in papaya fruit juice germinate poorly, and this shows the presence of inhibitors. Simple washing of seeds removes inhibitors to a certain extent and improves seed germination.

*Temperature.* Papaya seed germination is influenced by temperature, with an optimum temperature of around 35°C. Temperatures below 23°C and above 44°C are detrimental to germination. Night temperatures are more critical for seedling growth than day temperatures or day length. *Chemical and growth substances.* Endogenous growth promoters are found in the embryo and endosperm, while inhibitors are present in the sarcotesta. Seedling emergence improves with removal of the sarcotesta and also by soaking seeds in potassium nitrate (1.0 M), gibberellic acid (600 ppm), and thiourea. Higher concentrations of thiourea (4 to 10 grams per liter [g·liter –1]) affected germination and vigor. Seeds treated with gibberellic acid gave rapid emergence of seedlings and higher germination, particularly in seeds without sarcotesta.

*Soil salinity.* Seed germination is delayed and seedling growth is adversely affected by the presence of higher salt concentrations. Seedlings are slender, stunted, and cannot survive long under saline conditions beyond 4 millimhos per centimeter (mmhos/cm).

## Storage Conditions

Papaya seeds deteriorate rapidly at higher storage temperatures and relative humidity. They exhibit intermediate to recalcitrant and orthodox storage behavior. Fresh seeds give higher germination and seedling vigor that will decline on storage. The seed longevity is greatly influenced by type of container and storage conditions.

*Temperature.* Papaya seeds maintain viability for three months under ambient temperatures. Seed viability decreases at lower storage temperatures. Seeds stored at room temperature give higher germination than seeds stored in a desiccator or refrigerator. Singh and Singh reported that seeds remained viable for eight months at room temperature when packed in sealed polyethylene bags. Seed germination decreases with increase in storage period. Higher seed viability and vigor are maintained at 4 to 6°C than at room temperatures. The loss of viability was attributed to accumulation of inhibitors in seeds. Seeds stored at 10°C and 50 percent RH in cloth bags or at 5°C in sealed moisture-proof packages preserved viability reasonably well during six years of storage. High seed germination is recorded in seeds stored at 10°C after two years of storage. However, seed germination is affected when stored at 5°C and –18°C (Figure-2).

*Seed packaging.* Packaging is very important to protect the seeds from higher humidity and insects or pathogens. The selection of packaging depends on duration and conditions of storage. Loss of viability is greater in open storage than in sealed storage using polyethylene bags or glass jars. Seed viability can be preserved for 11 months when seeds are mixed with ash and stored in a capped blue bottle. Seed germination is higher (57 percent) in seeds stored in plastic bags at 4 to 6°C after 29 months of storage, and seeds stored in cloth bags at room temperature gave the lowest germination (16 percent). Seeds stored with nitrogen gas in laminated aluminum foil pouches exhibited higher germination during ambient storage. Seed germination was higher in seeds stored in paper bags, followed by seeds stored in polyethylene bags. Papaya seeds need air circulation during storage and can maintain viability for two years in paper bags at 10°C without significant loss in vigor.

*Seed storage with desiccants.* High seed moisture is injurious to storage life, and chemical desiccants are used to regulate moisture during storage. Seeds stored with desiccants maintain initial viability. Seed viability and vigor are preserved for four years in seeds stored with silica gel, as compared to one year without silica gel under ambient conditions (see Figure.3). Viability decreases to 50 percent in seeds without silica gel after nine months of storage.

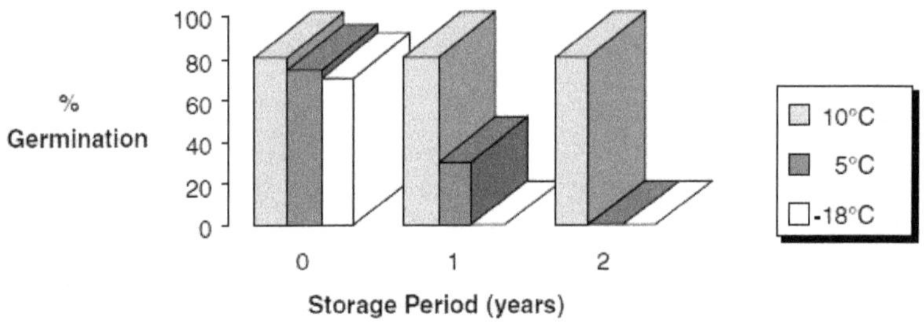


*Figure.2. Effect of Temperatures on Seed Viability in Papaya*

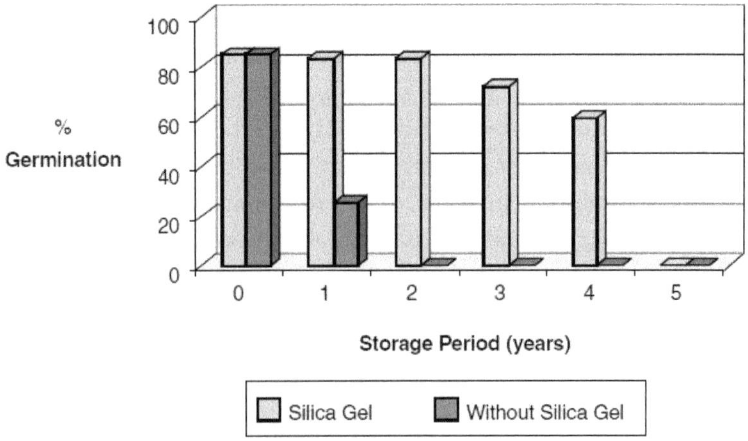


*Figure.3. Seed Storage with Silica Gel in Papaya*

# Chapter 5

# GRAPE: *VITIS VINIFERA* L.

## INTRODUCTION

Grape is a popular fruit crop in temperate regions. Its cultivation is expanding rapidly in the tropics and subtropics. It is grown for the juicy edible berries. Fruits are tasty and nourishing, containing high carbohydrates and minerals such as magnesium, potassium, phosphorus, and vitamins B1 and B2. Grape is easily digestible because it contains high amounts of glucose and fructose. It is also used in the preparation of juice, wine, syrup, and raisins.

### Origin and Distribution

Grape has its origin in Central Asia, particularly in Armenia, and wild relatives are grown in areas between the Black and Caspian Seas. Grape is predominantly cultivated in Afghanistan, Australia, Canada, England, France, Germany, India, Israel, Italy, the Philippines, Portugal, Russia, Spain, and the United States.

### Morphology

Grape belongs to family Vitaceae and has somatic chromosome number 2n = 38. It is a dicotyledonous, deciduous, and woody climbing vine. Leaves are broad and cordate, with three to five lobes and intermittent tendrils. Flowers are borne in panicle formation and are mostly perfect. Fruit cluster is variable in size, form, and color; the fruit skin adheres to the pulp. Seeds are brown, none to four in number, and pear-shaped.

## SEED STORAGE

Grape is propagated by cuttings, and seldom by seeds. However, seeds are used in evolving new cultivars through hybridization and in long-term conservation of genetic diversity through seed storage. Grape seeds show orthodox storage behavior, where drying or exposure to chilling temperatures does not kill seeds. Fresh seeds exhibit dormancy, which is beneficial for better seed storage.

### Seed Collections

Grape is a cool-season crop. It grows well in temperate regions, except in areas of extreme cold. It needs warm growing conditions during fruit development and ripening. Rooted cuttings are used for grape propagation. Vines are suitably trained on arches, trellises, or poles (see Figure.1). Flowering occurs on new growth. Fruits are nonclimatric; therefore, ripe bunches are plucked for seed purposes. Berries are crushed without injuring the seeds. Seeds are separated from the pulp, washed in water, dried, and packed according to storage requirements.

*Fruit maturity.* Seeds from mature berries emerge sooner and produce more plantlets of higher dry weight than seeds from green berries. Grape berries harvested at the green and mature stages and stored in paper bags at 20°C for 12 months showed seed germination percentage of 14.5 and 73.5, respectively.

### Seed Germination

Freshly extracted seeds normally contain a high proportion of dormant seeds. Scott and Ink  reported that fresh seeds had low germinability and low seedling vigor. Plants were dwarfs when such seeds were sown.

*Figure.1. Grapevine with Ripe Berries*

Seeds obtained from self-pollination are lower in viability than seeds from open pollination. Similarly, seeds obtained from grafted plants showed higher viability than seeds from selfrooted plants. Harmon and Weinberger noted that stratification or moist ripening results in higher seed germination. Grape seeds stored at 4 to 5°C for 60 days showed improved percentage of germination and early emergence , while Eris recommends storage for 120 days. Seed germination can also be hastened by keeping seeds in a peat-sand (1:1) mixture in polyethylene bags at 0 to 2°C. In another study, temperature regime (0 to 10°C) and storage duration (up to five months) increased germination. The most effective temperature for higher germination (40 percent) was 0°C. The effect decreased progressively with higher temperatures, and only 5 percent germinated at 10°C. However, a longer period of stratification at 1 to 3°C for 180 to 210 days improved germination. These seedlings grew vigorously and showed improved winterhardiness. Selim and colleagues reported that warm stratification (18°C for 30 or 60 days) followed by cold stratification (5°C for 30 days) gave higher germination and a higher number of normal seedlings.

The improvement was attributed to an increase in endogenous level of gibberellic acid (GA)-like substances. Furthermore, gibberellic acid treatment replaces the chilling or light requirement for germination. A long prechilling treatment is rarely successful in promoting germination of dormant seeds. Gibberellins are more effective in promoting germination of dormant seeds than other growth substances. Chohan and Dhillon reported that auxin activity increases with longer stratification. Although no gibberellin activity occurs in fresh seeds, it is higher in extracts of moist seeds stratified for 60 days. Abscisic acid-like inhibitor content is initially higher but is reduced on stratification. In many grape varieties, seeds are empty and do not germinate in spite of moist ripening or gibberellin applications.

*Heat treatment.* Grape seed coat is hard and takes longer to imbibe. Immersing seeds in hot water (27 to 54°C) for 24 h also does not break the dormancy.

*Light.* Seed exposure to continuous light promotes the germination process. Pereira and Maeda reported that seeds of cv. Patricia germinated well in continuous light for 48 h but failed in dark conditions.

## Storage Conditions

Grape seeds are short-lived under ambient conditions. Seeds deteriorate at higher temperature and relative humidity. Dolgova noted poor germination of seeds when stored in dry conditions for six to eight months.

Viability is preserved when seeds are stored in the berry itself or stratified in damp sand at 1 to 3°C for 180 to 210 days. Seed drying after harvest and/or stratification at 18.3 to 21.1°C reduces germination slightly, and even more through drying without subsequent stratification. Dry seeds stored at 4.4°C for three months showed higher germination than those stored at room temperatures. Seeds stored in paper or polyethylene bags at 20°C germinated poorly, unlike those stored at 2°C in polyethylene bags. Furthermore, fresh seeds stored in glass jars or paper bags remained viable at 10°C, and dry seeds, at 20°C. Seeds stored in polyethylene bags at 5 and –18°C for two years resulted in 45 and 41 percent germination, respectively, after gibberellic acid (2000 ppm) applications. Viability (67 percent) was extended further, for five years, by packing seeds in polyethylene bags and storing at 5°C.

# Chapter 6

# PINEAPPLE: *ANANAS COMOSUS* (L.) *MERR.*

## INTRODUCTION

Pineapple is a major tropical fruit cultivated in hot and humid regions for the edible fruits. Fruits are eaten fresh after peeling the rind. It is also canned in pieces or processed for jams, juices, and squashes. Fruits are a rich source of vitamins A, B, and C and of minerals such as calcium, phosphorus, and iron. They are a good source of the protease enzyme bromelain, which is used in medicine and in the food-processing industry.

### Origin and Distribution

Pineapple originated in Brazil. Many primitive and wild relatives are found in the Caribbean, Central America, and northern South America. Its cultivation later spread to other countries, such as Australia, China, Colombia, Cuba, Ghana, India, Indonesia, Japan, Kenya, Malaysia, Mexico, Mozambique, Nigeria, the Philippines, South Africa, Taiwan, Thailand, and the United States.

### Morphology

Pineapple belongs to family Bromeliaceae and has chromosome number 2n = 50. The plant, a perennial and monocotyledonous herb, grows up to 1 m in height. Leaves are narrow, thick, rosette shaped, with a pointed tip and spiny margin, and are about 60 to 80 in number. The inflorescence is compact and bears 100 to 200 flowers. Flowers are perfect, opening first at the basal portion and later moving upward to produce fruit parthenocarpically.

Fruit is cylindrical and dark green, changing to yellow or reddish orange on ripening. Fruit is seedless and has yellow flesh. Artificial pollination produces 2,000 to 3,000 seeds. Seeds are brown and rough; embryos are small with hard endosperm.

## SEED STORAGE

Pineapple is commonly propagated by vegetative means. Shoots from crown, slips, and suckers are used for propagation. Seeds are used in evolving new cultivars through hybridization. Pineapple seeds show orthodox storage behavior. They can withstand desiccation, and longevity is extended at lower-temperature storage. Many cultivars do not produce seeds, and fruit develops parthenocarpically. In certain cases, seeds are produced through artificial pollination. These can be used to a certain extent for conservation of genetic diversity.

### Seed Collections

Pineapple is a warm-season crop. It needs hot and dry conditions for rapid growth. Fruit takes longer to grow at higher latitudes, and fruit quality improves at lower temperatures. It requires deep, fertile, and sandy loam soils. Plants take longer (four years) to bear fruit when propagated by seeds than when grown from suckers (18 months). Fruit becomes yellow to reddish orange on ripening. Seed sets on artificial pollination and takes about five to six months for maturation. Seeds are present inside the carpel about 1.5 cm deep and are extracted by cutting fruit longitudinally. These are washed, dried, and used for storage.

### Seed Germination

Seeds have a very tough seed coat and hard flinty endosperm. Therefore, they germinate slowly and unevenly. Normally, it takes ten days for germination after acid scarification at 30°C. Seed incubation at 30 to 35°C and scarification with concentrated sulfuric acid for 30 min is beneficial for increased germination. Iyer, Singh, and Subramanyam reported faster and higher germination after exposing seeds to intermittent mist (for 20 seconds [sec]/ 3.5 min intervals). This is attributed to softening of the hard seed coat or leaching of water-soluble inhibitors by the water spray.

### Storage Conditions

Pineapple seeds are orthodox in storage behavior. They store better under cool, dry conditions. Well-dried seeds are suitably packed in moisture- proof containers and stored at low temperature for longer storage.

# Chapter 7

# GUAVA: *PSIDIUM GUAJAVA* L.

## INTRODUCTION

Guava is an important fruit crop in the tropics and subtropics. Mature green or ripe fruits are consumed fresh and used in preparation of products such as juices, pastes, jams, jellies, and pies. Fruits are rich in pectin and vitamin C, and red-fleshed cultivars are high in carotene content. Leaves and barks contain a large amount of tannin and are used in the dyeing and tanning industry. Leaves have medicinal properties and are used in controlling diarrhea and stomach ailments.

### Origin and Distribution

Guava's wild relatives and primitive types are found in Brazil, Mexico, and Peru. Later its cultivation spread to other African and Asian countries. Guava is cultivated predominantly in Algeria, Australia, Brazil, China, Colombia, Cuba, Egypt, India, Israel, Mexico, Peru, the Philippines, South Africa, the United States, and the West Indies.

### Morphology

Guava belongs to family Myrtaceae and has chromosome number 2n = 22. It is a small, dicotyledonous, semi-deciduous tree growing up to 3 to 10min height. Foliage is thick and dark green. Leaves are opposite and oblong with a rough surface. Flowers are white, with one to three in a cluster, axillary, hermaphrodite, and fragrant. Calyx is entire in bud and splits into four to six lobes; petals number four to five; stamens are numerous; the ovary has four to five locules. Fruit is a berry, globular or pear-shaped; the exocarp is pale green to bright yellow; the mesocarp is white, yellow, pink, or red and fleshy and is embedded with many seeds. Seeds are small (3 to 5 mm), hard, reniform, and yellowish in color.

## SEED STORAGE

Guava is propagated both sexually and asexually. Many existing orchards have been raised from seedlings. Guava seeds show orthodox storage behavior. They preserve high viability and vigor under dry, cool conditions. Seeds can be used for long-term conservation of genetic diversity in the gene bank. Seeds are also involved in crop improvement through hybridization programs. Guava seeds can be stored well at low moisture and low temperature without significant loss of viability.

### Seed Collections

Guava is cultivated in a wide range of climatic and soil conditions. It prefers warm and dry conditions for optimum growth. Frost and temperatures below 16°C affect plant growth. It grows well in deep, friable, well-drained soils and is tolerant to saline conditions. Seeds and air layers are used for propagation. Air layer plants do not show variations and retain the identity of the mother plant. Tree commences flowering from the second year onward.

Both self- and cross-pollination by insects occur. Fruit takes about five months to mature following anthesis. Healthy and ripe fruits are selected for seed purposes; seeds are separated from the pulp, washed thoroughly, dried in shade, and packed suitably for storage.

### Seed Germination

Guava seeds germinate poorly and unevenly and require more time for seedling emergence. Teaotia and Singh reported a variation in germination percentage from 22 to 72 in different cultivars. Poor germination of guava seeds is primarily attributed to dormancy and to hard testae. Seed scarification, either mechanically or chemically with acids, overcomes delay and the slow germination process. Seed soaking in tepid water before sowing increases the germination percentage, and use of cold water is safer than using hot water. Seed soaking in cold water for 12 h or in boiling water for 5 min is beneficial for higher germination. Seed soaking in concentrated hydrochloric acid for 3 min or in concentrated sulfuric acid for 5 min eliminates the dormancy. Further soaking of seeds in ethephon (5,000 ppm) increased seedling emergence and encouraged better vegetative growth of plant. Exposure of seeds to alternate temper atures of 20/30°C for 16/8 h, respectively, improved seed germination.

## Storage Conditions

Guava seeds show orthodox storage behavior. Thus, well-dried seeds appear to preserve longer at low temperature. These seeds retain viability for about one year under ambient conditions. According to Chacko and Singh , seed viability reduced significantly after 18 months of storage when stored with or without charcoal in open or sealed containers.

# Chapter 8

# BANANA (*MUSA* SPP.)

## INTRODUCTION

Banana is a leading fruit crop of the tropics. It is cultivated for its edible fruits. Ripe fruits are consumed fresh, while unripe fruits are cooked as a vegetable. Almost all parts of banana plants are used in one form or another, such as for wrappers, fibers, in medicine, in religious functions, as cattle feed, etc. Bananas are a rich source of energy, and they contain high amounts of carbohydrates, calcium, phosphorus, iron, magnesium, sodium, potassium, and vitamins A, B, and C.

### Origin and Distribution

Banana originated in Southeast Asia. The cultivated banana is derived from *Musa acuminata* (A genome) and *M. bulbisiana* (B genome) types. Banana grows well in humid tropical lowlands and is predominantly distributed between 30°N and 30°S of the equator. Major banana-growing countries are Australia, Bangladesh, Brazil, Burundi, Colombia, Cuba, Ecuador, Egypt, Fiji, Ghana, Guatemala, India, Indonesia, Israel, the Ivory Coast, Kenya, Malaysia, Mexico, Panama, the Philippines, South Africa, Tanzania, Uganda, the United States, Venezuela, the West Indies, and Zaire.

### Morphology

Banana belongs to family Musaceae and has chromosome numbers 2n = 22, 33, and 44. Seed sets in wild diploid banana but fails in triploids due to sterility. Banana is a perennial, monocotyledonous, robust herb. The corm, or underground stem, has short internodes. It takes 7 to 9 months for flowering after planting suckers. The inflorescence is a spike that produces 12 to 20 flowers per node. The first 5 to15 basal nodes produce female flowers, while the terminal node forms male flowers (see Figure.1). Fruit is a berry and takes 90 days for maturity from anthesis. Seeds are subglobose or angular, 5mmin diameter, very hard, and endospermic, with a minute embryo at micropylar end.

*Figure.1. Banana Fruit Bunch Supported by a Pole*

## SEED STORAGE

Banana is propagated by both sexual and asexual methods. Seeds are used in hybridization for evolving new cultivars, while suckers are planted for raising commercial orchards. Banana seeds show orthodox storage behavior, retaining high viability on drying and conservation at chilling temperatures.

### Seed Collections

Banana is a warm-season crop. It requires a hot, humid climate for optimum growth. Heavy winds and hailstorms damage the crop. Plant needs support, especially during fruit development. It is a shallow-rooted plant and requires fertile, well-drained soil. It is a cross-pollinated crop, and pollination is achieved by ants, bees, wasps, and bats. Fruit matures in 11 to 16 months after planting, depending on cultivars and growing conditions. On maturity, bunches are harvested and fruits allowed to ripen. Seeds are extracted from the pulp, washed, dried, and packed for storage.

### Seed Germination

Banana seeds germinate readily after extraction. However, they become dormant on drying. Seed germination is hypogeal. Seeds germinate with wa- ter imbibitions, and a micropylar plug extrudes, followed by the radicle, which is short-lived and quickly replaced by seminal adventitious roots. Then, the plumule grows and becomes a normal seedling. Seed germination is higher in seeds extracted from matured fruits than in those from immature fruits. Low and erratic germination were attributed to degree of seed maturity. Seed viability is decided by climatic conditions, period of pollination, and position of seeds in fruit. Banana seeds germinate between temperatures of 10 to 37°C, and germination is affected beyond this range of temperature. Dormant seeds do not germinate in this temperature regime. However, they do germinate readily at alternate temperatures of 15/35°C. Chipping imbibed seeds gives rapid germination.

Nondormant seeds germinated at 25 to 35°C within three weeks of sowing. Storage of *Musa bulbisiana* seeds in a desiccator at room temperature for three months resulted in an increase in germination from 73 to 95 percent. Seed germination can be enhanced by six months' storage in 5 percent carbon dioxide atmosphere.

## Storage Conditions

Banana seeds are orthodox in storage behavior and germinate readily. Seeds withstand desiccation and their longevity is extended at low temperature. Purseglove reported that seeds buried in soil remained dormant for more than one year under dry conditions. Dried seeds remain viable for a few months to two years. Seeds are stored suitably in moisture- proof containers, such as thick polyethylene bags, and in laminated aluminum foil pouches at low or subzero (–20°C) temperatures, or even at cryogenic temperatures, for longer conservation of viability and vigor.

# Chapter 9

# JACKFRUIT: *ARTOCARPUS HETEROPHYLLUS* LAM.

## INTRODUCTION

Jackfruit is largely grown in the tropics and subtropics. Normally, a few trees are planted in orchards and home gardens. Fruits are appreciated for their sweet crisp or soft pulp and characteristic flavor. Immature fruits are cooked as vegetables. Pulp of ripe fruits is eaten fresh, cooked, or preserved in syrup. Seeds are boiled, roasted, and eaten. The green foliage and rind are used as cattle feed, and the wood from the plant forms an excellent timber. The leaves also possess medicinal properties, and the fruits are a rich source of iron, phosphorus, calcium, and vitamin A.

### Origin and Distribution

Jackfruit originated in India. The wild relatives are found in western parts of India. Later, it was introduced to other countries in Africa and Asia. Jackfruit is cultivated in Brazil, India, Malaysia, Myanmar, the Philippines, Sri Lanka, Thailand, and the West Indies.

### Morphology

Jackfruit is a member of the Moraceae family and has chromosome number 2n = 56. It is an evergreen, monoecious, dicotyledonous large tree that grows up to 20 m in height. All parts of the plant produce white latex. Foliage is dark green; leaves are alternate and ovate, with shiny upper surfaces.

Male and female inflorescences are borne separately, and the male inflorescence is axillary on the leaf twigs, 15 to 30 centimeters (cm) long, drooping, green, and club-shaped. The female inflorescence forms on short stout twigs on the trunk and thicker branches, and it is oblong in shape. Fruit is a large syncarp, dark green, pear-shaped, and glabrous with fleshy spines, and it contains many seeds. Pulp is yellow and waxy. Seeds are large and oblong with a thick gelatinous covering.

## SEED STORAGE

Jackfruit is commonly propagated from seeds sown directly in the field or from transplanted seedlings. Fruits contain normally around 100 seeds in table types and 500 seeds in inferior types. Seeds weighing between 3 to 6 g are considered good for planting, and those less than 3 g are not suited for propagation. Seeds must be sown immediately after extraction from fruit, or they lose their viability. Jackfruit seeds exhibit recalcitrant storage behavior, where seeds are killed on drying and exposure to chilling temperatures. Such seeds are difficult to store for long periods. However, seeds can be stored for a few weeks to few months with available storage methods.

### Seed Collections

Jackfruit grows well in hot, humid climates. It prefers deep, fertile, and well-drained soil. Seeds, which are predominantly used in propagation of jackfruit, germinate up to 15 days after extraction, and thereafter germination declines. The flowers are pollinated by insects and take 100 to 120 days for fruit maturity. In general, a tree bears 200 to 300 fruits (see Figure.1).

Fruits are harvested at maturity and stored for a few days for ripening. They emit a peculiar odor on ripening. Large, healthy, and normally shaped fruits are selected for their seeds. Seeds are extracted and washed, with large and bold seeds selected for storage.

### Seed Germination

Fresh seeds germinate readily and take three to eight weeks to complete the germination. Dry and shriveled cotyledonous seeds do not germinate. Viviparous seed germination is reported in early maturing types (25 to 36 percent) and in late-maturing types (37 to 39 percent). Seed germination is higher at a constant temperature range of 25 to 30°C and in the presence of light. It is improved further on seed soaking in gibberellic acid (100 ppm) and naphthaleneacetic acid (25 ppm).

### Storage Conditions

Jackfruit seeds are short-lived under ambient conditions. Their vigor decreases on increase in storage period. They lose viability and vigor rapidly with decrease in seed moisture content. High-moisture seeds failed to preserve viability at low temperature and showed chilling injury and loss of viability. Fresh seeds gave rapid and higher germination and better seedling survival than sowing the seeds stored at 6°C for 15 days. Jackfruit seeds stored at ambient temperatures and at 6°C germinated well, in the range of 80 to 86 percent, after 22 days of storage; none of the seeds germinated after 38 days of storage. Moist storage in a screw cap bottle at 16 to18°C is an ideal method that preserves seed viability for 45 days. Tang and Fu preserved high seed viability and vigor for more than two years by packing washed seeds in perforated polyethylene bags and storing them at 15°C. Seed storage at 5 and 10°C showed chilling injury after 40 and 300 days of storage, respectively.

*Figure.1. Jackfruit Tree with Many Fruits*

# Chapter 10

# POMEGRANATE: *PUNICA GRANATUM* L.

## INTRODUCTION

Pomegranate is a popular dessert fruit in the tropics and subtropics. It is extensively cultivated in the Mediterranean region for the fruit's juicy pulp. It is also grown in home gardens as an ornamental plant. The fruit is processed for juice, syrup, and jelly preparations. Seeds are used in preparation of several food dishes. It is rich in minerals such as calcium, iron, phosphorus, and vitamins B and C. The bark from the plant is used for medicinal purposes, such as curing stomach ailments and diarrhea, and the seeds show antibacterial properties.

### Origin and Distribution

Pomegranate is native to Iran. Its wild types and landraces are grown in Central Asia, including in Afghanistan, Iran, and Pakistan. Pomegranate is cultivated in Afghanistan, Bulgaria, China, Egypt, India, Indonesia, Iran, Israel, Italy, Japan, Morocco, Myanmar, Pakistan, Russia, Spain, Syria, Tunisia, Turkey, and the United States.

### Morphology

Pomegranate belongs to family Punicaceae and has chromosome number 2n = 16. It is a small, bushy, dicotyledonous tree that reaches 2 to 4 m in height. Leaves are opposite, dark green, glossy, and deciduous in cooler climates.

Flowers are showy, orange-red, and borne in clusters; the calyx is campanulate with five to seven lobes; petals number five to seven and are wrinkled; stamens are numerous; the ovary has three to seven cells. Fruit is a berry, leathery skinned, globose or spherical, brownish yellow to red, and surrounded by persistent calyx. Interior of fruit is composed of chambers that are filled with numerous seeds. Seeds are about 1 to 1.5 mm long and are surrounded by pink juicy pulp.

## SEED STORAGE

Earlier, many pomegranate orchards were raised from seedlings. These plants required a longer period to bear fruit and showed variations in fruit qualities. Of late, vegetative propagation by hard cutting of the wood has become a common practice in pomegranate cultivation. It gives uniform, early bearing, and true-to-type plants. However, seedlings are used in developing new cultivars through hybridization, and seeds are used in long-term conservation of genetic diversity in the gene bank. Seeds show orthodox storage behavior, withstand drying, and are capable of preserving viability at low temperatures.

### Seed Collections

Pomegranate grows better in semiarid climates, where cool winters and hot dry summers are prevalent. It fruits poorly under very humid conditions and can withstand drought conditions. It needs a dry hot climate during fruit development for high-quality fruit. Deep loamy soils are ideal for cultivation.

Plants flower throughout the year. Fruits take six months to ripe after anthesis. Self- and cross-pollination occur, the latter mainly by beetles. Plants commence bearing from the fourth year onward (see Figure.1).

Healthy and mature fruits are selected for seed purposes. Seeds are extracted from fruit, washed in water, and surface dried. The fresh seeds contain about 70 to 75 percent moisture, which is reduced to 5 to 7 percent for storage.

### Seed Germination

Pomegranate seeds germinate rapidly at constant temperatures of 30 to 35°C. Seeds germinated at 35°C give low-vigor seedlings. Seed exposure to alternate temperatures of 20/35°C for 16/8 h, respectively, promotes higher germination. The germination process is complete 28 days after sowing. Seed germination is faster within 14 to 30 days of sowing when seeds are exposed to prechilling, 1 to 5°C for 30 days. According to Cervelli , removal of the fleshy seed coat improves seedling emergence by 5 percent, and subsequent washing in water for 48 h increases it further, by 26 to 62 percent.

## Storage Conditions

In pomegranate, the pericarp is woody and leathery, loses moisture slowly, and protects seeds from insects and diseases. Mature fruits are stored for five to six months in cool conditions. Subsequently, seeds are extracted and used for further storage or for propagation.

*Figure 10.1. Early Fruiting in Pomegranate Tree*

Seed viability is preserved well in cool, dry conditions. Fruits stored for a shorter period retain seed viability and vigor. Riley recommended drying of seeds to 70 percent of their harvest weight and storing them at 4 to 5°C to preserve seed viability for three years.

# Chapter 11

# AVOCADO: *PERSEA AMERICANA* MILL.

## INTRODUCTION

Avocado is an important tropical fruit having high nutrient contents and the highest energy value among fruits. Fruits are normally consumed fresh along with sugar, cocoa, and other flavoring compounds. They are rich in unsaturated fats, proteins, minerals, and vitamins A, C, G, and thiamine. Avocado fruits, which are available throughout the year, have a nutlike flavor. Seeds also contain high amounts of oil and are used for cosmetic and medicinal purposes.

### Origin and Distribution

Avocado originated in Central America. Later it spread to other tropical and subtropical regions. It is cultivated in Australia, Brazil, Chile, Cuba, Ecuador, Haiti, Indonesia, Israel, Kenya, Madagascar, Mexico, Peru, South Africa, Spain, Thailand, and the United States.

### Morphology

Avocado belongs to family Lauraceae and has chromosome number 2n = 24. Avocado is an evergreen tree with a short trunk, large branches, and a shallow root system. Leaves are simple, ovate, and spirally arranged. Inflorescence is compact and axillary with numerous yellow flowers. Flowers are perfect, protogynous, and fragrant. Fruit is a fleshy berry, pyriform or globose, and yellowish green to purple in color. Flesh has a butterlike taste and is yellow in color. Seeds are large, globose, single, and whitish or pink in color.

## SEED STORAGE

Avocado is commonly propagated by vegetative means due to the heterogeneous nature of seeds. Many existing avocado orchards have been raised through seedlings. Seeds are sown immediately after extraction for raising rootstocks. These are short-lived and remain viable for two to three weeks after removal from fruit. Avocado seeds appear to be recalcitrant; they lose viability on desiccation and are susceptible to chilling temperatures.

### Seed Collections

Avocado grows well in tropical conditions. Some cultivars are susceptible to cold and are damaged by strong winds. It grows well in deep, fertile, sandy loam soil. It is propagated by grafting. Plants show a biennial bearing habit. Avocado is a cross-pollinated crop, and pollination occurs mainly by bees. The percentage of fruit set is low. Healthy and bold seeds are separated from the fruit, washed, and used immediately or processed for shortterm storage.

### Seed Germination

Seed germination is hypogeal and higher at a constant temperature of 25 to 30°C. Seedling emergence is slow and nonuniform due to the hard seed coat. Removal of the seed coat promotes early germination. Halma and Frolich reported that cutting a small portion off both ends of the seed gives quicker and even germination. Later, Johnston and Frolich suggested that cutting a slice off the tip and base of the seeds gives higher and uniform germination. Seed soaking in gibberellic acid (125 to 1,000 ppm) resulted in higher germination and better seedling growth.

### Storage Conditions

Avocado seeds are recalcitrant in storage behavior and remain viable for two to three weeks under ambient conditions. Seed viability can be maintained for several months by preserving seeds in sand or sawdust at 4.4 to 7.2°C , and for eight months in dry peat moss at 5.5°C. Thin polyethylene (0.025 mm) bags are well suited for storage because they allow gaseous exchange and preserve viability comparatively longer. However, seeds stored in polyethylene bags with holes invite *Penicillium* or *Aspergillus* fungal species. Brokaw noted that periodic exchange of air in plastic bags and the use of thinner packaging help maintain high viability (90 percent) for six months. Similarly, Spalding, Knight, and Reeder preserved avocado seeds for five months at 4.4°C when sealed in polyethylene bags along with fungicides.

# Chapter 12

# LITCHI: *LITCHI CHINENSIS SONN.*

## INTRODUCTION

Litchi is a popular dessert fruit in Southeast Asia. The translucent juicy aril of the fruit is eaten fresh, canned, preserved in syrup, dried, or frozen. Litchi fruit dried and used as a nut is popular in China. Fruit contains high amounts of carbohydrates, calcium, phosphorus, and vitamin B. Seeds are used for medicinal purposes.

### Origin and Distribution

Litchi belongs to family Sapindaceae and has chromosome number 2n = 28. Litchi originated in southern China. Later its cultivation spread to other subtropical regions. It is largely cultivated in Australia, Bangladesh, Brazil, China, India, Indonesia, Israel, Japan, Madagascar, Mauritius, Mexico, New Zealand, Seychelles, South Africa, Taiwan, Thailand, the United States, Vietnam, and the West Indies.

### Morphology

Litchi is an evergreen, dense, dicotyledonous tree that grows up to 10 m in height. Leaves are glossy, green, and pinnate with two to nine leaflets. Flowers are small, green, or yellow, in terminal panicles, with male and female flowers. Fruits are globose or oblong, borne in loose clusters with long peduncles, and the rind is dark red or yellow, brittle, and covered with sharp tubercles. Seeds are dark brown and covered with a pearl-white, translucent, juicy aril (see Figure.1).

## SEED STORAGE

Litchi is propagated sexually and asexually. Seeds do not produce trueto- type plants and take longer to bear fruit, which is often of inferior quality.,However, many old existing orchards have been raised from seedlings.

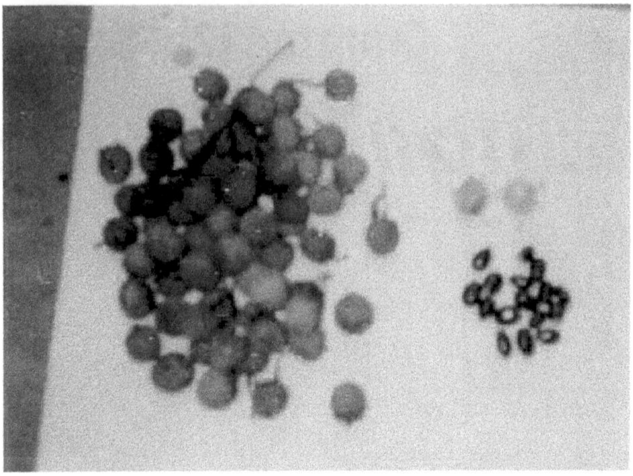

*Figure.1. Litchi Fruits and Dark Recalcitrant Seeds*

Presently air layering is a popular method for multiplication of litchi plants. Such plants bear fruit within four to six years of planting. Seeds are commonly used in raising rootstocks and in evolving new cultivars through hybridization. Litchi seeds show recalcitrant storage behavior and remain viable for four to six days at room temperature. Seed loses viability rapidly on drying and is killed by exposure to chilling temperatures.

### Seed Collections

Litchi grows well in subtropical conditions, with high summer and low winter temperatures, and in the tropics, particularly at higher altitudes. Low humidity and dry winds cause fruit splitting. Litchi prefers deep, welldrained soil. Trees raised from air layering take four to six years to yield, as compared to about twenty years for those raised from seedlings. Self- and cross-pollination occur, and flies, wasps, ants, and bees cause pollination. Fruits take about two months to mature and change their color on maturity. Fruits are allowed to ripen on the tree. Large and healthy fruits are selected for seed purposes. Seed loses viability immediately after extraction; therefore, it is planted immediately or processed for short-term storage.

### Seed Germination

Seeds germinate immediately after extraction. Seed germination is poor and er- ratic. Seedling emergence is relatively slow and nonuniform due to the hard seed coat.

Seed germination is higher at constant temperatures of 25 to 30°C, and it takes 16 days to complete. Seed germination and seedling emergence is rapid in large seeds. Seeds contain high amounts of starch, and its breakdown precedes germination. Starch degradation is a characteristic feature of germination.

Amylase is active at the initiation of germination, and retention of high amounts of starch in cotyledons denotes its partial breakdown, resulting in poor mobilization to growing seedlings.

*Fruit maturity.* Seeds develop rapidly immediately after fertilization, which is followed by the formation of the aril and membranous mesocarp. Seed germination improves with advance in maturity stage. Fruit color changes from yellow-green to dark reddish brown on maturity. Seed attains maximum dry weight eight weeks after anthesis and gives 42 percent germination. Highest seed germination (100 percent) was recorded in fruits during the ninth week after anthesis.

*Seed drying.* Litchi seeds are susceptible to desiccation injury. They lose germination capacity and vigor potential on drying. The seed coat and cotyledon turn brown on drying after four to six days of storage at room temperature. Initial seed moisture content is 28.5 percent and viability is 100 percent immediately after seed extraction. Seed viability is lost completely if moisture drops to 19 percent. It appears that 20 percent moisture is critical for litchi seed storage; below this level seed viability is injured. During this period, peroxidase content increases and ascorbic acid decreases.

## Storage Conditions

Litchi seeds lose viability rapidly on detachment from fruit and are recalcitrant in storage behavior. They can be stored for a few weeks to a few months with prevailing storage methods. King and Roberts reported that such factors as drying injury, chilling damage, microbial contamination, and germination of high-moisture seeds contributed to loss of viability in litchi. Suitable storage methods need to be developed accordingly to meet these requirements in extending seed longevity.

*Seed storage in fruit.* Fruit storage is a simple method and is suitable for short-term maintenance of seed viability. The fruit's rough, thick skin slows loss of moisture and keeps the fruit in good condition. Ray and Sharma reported that seeds retained in fruit treated with benomyl (0.05 percent) and wax emulsion (6 percent) and sealed in polyethylene bags maintained 42 percent viability up to 24 days of storage.

*Seed storage in water.* Rapid loss of moisture is the main cause for poor storage of litchi seeds. Seeds stored in water gave 75 and 30 percent germination after 7 and 15 days of storage, respectively.

*Seed storage in gases.* Singh and Prasad observed rapid loss of seed viability both at ambient temperatures and in the refrigerator. Further, storage in moisture-proof containers caused seed sprouting. Seeds stored in the open remained viable (15 percent) for five days. Seeds stored in sealed polyethylene bags gave 50.7 percent germination after ten days. Sowa, Roos, and Zee proposed a new approach to recalcitrant seed storage as in a combination of nitrous oxide ($N_2O$) and oxygen ($O_2$). However, seed storage in only anesthetic gas is not beneficial. When combined with oxygen (80 percent $N_2O$+ 20 percent $O_2$) it resulted in 92 percent germination after three months of storage at 8 to 10°C, compared to 44 percent germination in seeds stored in air.

*Seed storage with chemicals.* Seed pelletization is useful in reducing moisture loss. Seeds soaked in calcium chloride (4 percent) followed by pelletization with sodium alginate (4 percent) preserved viability (53 percent) for 80 days at 10°C.

# Chapter 13

# LOQUAT: *ERIBOTRYA JAPONICA* LINDL.

## INTRODUCTION

Loquat is a popular dessert fruit in the subtropics. Fruit is eaten fresh and processed for jam, juice, squash, and jelly. Fruit contains high amounts of carbohydrates, phosphorus, calcium, iron, and vitamin A.

### Origin and Distribution

Loquat originated in China. It is cultivated predominantly in Australia, eastern Asia, South Africa, South America, and the Mediterranean region. Some major loquat-producing countries are Australia, China, Italy, Japan, and the United States.

### Morphology

Loquat belongs to family Rosaceae and has chromosome number 2n = 34. It is an evergreen, dicotyledonous, dense tree that grows up to 7 m in height. Leaves are large, alternate, lanceolate to obviate, with a shiny upper surface. Flowers are white, fragrant, and borne on terminal panicles; the calyx is five lobed; petals number five; stamens number twenty; and pistil five joined toward the base. Fruits are borne in cluster; they are round, oval, or pyriform and yellow to orange in color. Flesh is deep orange and juicy. Seeds are large, smooth, brown, and oval; they normally number three to five in a fruit, rarely up to ten.

## SEED STORAGE

Loquat is propagated by vegetative means, such as air layers and in-arch grafting. Seed germinates readily, but seedling trees bear inferior fruits that are not true to type. Seeds are used for raising rootstocks and for evolving new cultivars through hybridization. Loquat seeds show recalcitrant storage behavior and do not withstand loss of moisture and exposure to chilling temperatures during storage.

### Seed Collections

Loquat grows well in mild subtropical climates. Temperatures below the freezing point are harmful. Trees are resistant to heat and drought conditions, preferring well-drained soil. Rootstocks are commonly raised from seeds. Self-incompatibility occurs in many loquat cultivars and invites cross-pollination. Trees bear fruit from the third year onward. Fruits are formed in clusters and harvested when all are mature. Large, uniform, and healthy fruits are selected for seed purposes. Seeds are removed from fruits, washed, and prepared for storage.

### Seed Germination

Seeds require 21 days for germination. Seedling emergence is rapid, uniform, and higher when seeds are exposed to chilling at 1 to 5°C for 30 to 60 days.

### Storage Conditions

Loquat seeds are short-lived under ambient conditions. Seeds deteriorate rapidly after one month of storage at 5°C. Seeds stored in moist conditions at 5°C and high relative humidity gave 92 percent germination after six months of storage.

# Chapter 14

# LONGAN: *EUPHORIA LONGAN STEUD.*

## INTRODUCTION

Longan is also called cat's eyes. It is cultivated widely in Southeast Asia. Fruit is sweet and tasty and the juicy aril is eaten fresh. Fruit is also dried, canned, or frozen. The tea made from longan fruit is considered a tonic. Dried flowers are used for medicinal purposes, and the wood is a valuable timber for furniture. Seeds have a high saponin content and are used in manufacturing shampoo.

### Origin and Distribution

Longan is native to China. It is cultivated in Australia, China, India, Indonesia, Java, Malaysia, the Philippines, Sri Lanka, Thailand, the United States, and Vietnam.

### Morphology

Longan belongs to family Sapindaceae and has chromosome number 2n = 30. Trees are evergreen, dense, and grow up to a height of 40 m. The inflorescence is terminal; flowers are unisexual, yellowish-brown, and selfincompatible. Fruits are globular and smooth and contain black seeds.

### Seed Storage

Longan is propagated by seeds and grafting. Seeds remain viable for a short period under ambient conditions. Longan seeds are recalcitrant in storage behavior and lose viability on drying and preserving at chilling temperatures.

## Seed Collections

The longan tree grows well in cool climates and at an altitude up to 1,000 m. Seedlings and in-arch grafts are used for raising orchards. It is self incompatible, and bees bring about pollination. Fruit requires five months from flowering to harvest. Ripe and healthy fruits are selected for seed purposes. Seeds are removed, washed, and used immediately for propagation or for short-term storage.

## Seed Germination

Seed germination and seedling development is slow and nonuniform due to the hard seed coat. Seedling emergence is faster and higher at 25 to 30°C in presence of light. Germination takes about 17 days to complete.

## Storage Conditions

Longan seeds are short-lived and difficult to preserve for long periods. Chen and Fu reported that seeds remain viable for four to six days at room temperature. Seeds experience rapid moisture loss, a decrease in ascorbic acid, and an increase in peroxidase content.

*Seed storage in fruit.* Fruit storage is a simple method for seed storage. Fruits treated with benomyl (0.05 percent) stored in polyethylene bags retained viability for five days. Further, longevity is extended with refrigeration.

*Seed storage with gases.* Seed storage with a combination of nitrous oxide and oxygen preserved viability relatively longer. Sowa, Roos, and Zee reported that seeds stored in 80 percent N2O + 20 percent O2 remained viable (70 percent) for seven weeks, as compared to nil with air storage.

# Chapter 15

# MANGOSTEEN: *GARCINIA MANGOSTANA* L.

## INTRODUCTION

Mangosteen is a popular delicious fruit in Southeast Asia. Pulp is eaten fresh, canned, and made into squash, syrup, or jelly. It is also used for medicinal purposes. The wood is valued for high-quality timber in the preparation of a wide range of furniture.

### Origin and Distribution

Mangosteen is native to Malaysia. The wild species of mangosteen are found in Malaysia and India. Its cultivation was initially done in peninsular Malaysia. Later it spread to other tropical countries, but with limited success (gave low yield and poor fruit quality). Mangosteen is grown in Costa Rica, Ecuador, Honduras, India, Indonesia, Kampuchea, Madagascar, Malaysia, Myanmar, Panama, the Philippines, Sri Lanka, Thailand, and Vietnam.

### Morphology

Mangosteen belongs to family Guttiferae and has chromosome number 2n = 76. An evergreen tree is glabrous, dioecious, and slow growing, reaching a height up to 15 m. Leaves are opposite and ovate. Flowers are unisexual, yellow, and borne terminally; sepals number four; petals number four, and the ovary is four to eight celled. Fruit is a subglobose berry with persistent calyx and stigma lobes; pericarp is purple and tough with a yellowish resin and a specific flavor. Seeds vary in number from none to three and are formed from nucellar tissue.

## Seed Storage

Vegetative propagation is seldom successful in this crop; hence, seeds are used for raising plants. Seeds have low and short viability after removal from fruit. Mangosteen seeds show recalcitrant storage behavior. Seeds are killed on desiccation and by exposure to chilling temperatures.

## Seed Collections

Mangosteen grows well in hot, humid climates where temperatures range from 25 to 35°C and the humidity is higher than 80 percent. Mangosteen adapts to high rainfall conditions. Trees raised from seedlings are true to type, as seeds are formed parthenocarpically from nucellar tissue. Plants bear fruit 10 to 15 years after planting. Healthy and ripe fruits are collected for seed purposes. Seeds are difficult to preserve, even for a week. They are therefore used immediately upon removal from fruit.

## Seed Germination

Mangosteen seeds germinate readily and do not show any dormancy. Seed germination is rapid and higher at a constant temperature range of 25 to 35°C in the presence of continuous light, and it is completed in 28 days.

## Storage Conditions

Mangosteen seeds are recalcitrant in storage behavior and lose viability quickly when the thin membrane around the seed is damaged and subjected to drying or low storage temperatures. Seed viability is lost by four weeks after air-drying and by one week in a desiccators. Seeds stored in fruit show delayed germination.

Mangosteen seeds along with fruit stored better at 25°C than at 15°C or 35°C. Seeds stored in a humid atmosphere at 25°C maintained viability (56 percent) for seven weeks. Moist or dry seeds did not germinate after one week of storage at 10°C. However, seeds stored in moistened charcoal at room temperature gave highest percentage of germination. Cox observed that seed storage in moist charcoal or moss at room temperature preserves viability for a few months.

# Chapter 16

# PASSION FRUIT: *PASSIFLORA EDULIS SIMS*

## INTRODUCTION

Passionfruit is widely cultivated in the tropics and subtropics in small orchards or in home gardens. The pulp and seeds are eaten fresh from shell and used in salad. The flavored pulp is used in the preparation of jams, jellies, and beverages. Seeds contain a high amount (20 percent) of edible oil.

### Origin and Distribution

Passionfruit originated in South America, where wild types are abundant. The two common types of passionfruit are purple and yellow. Purple passionfruit is native to southern Brazil. The yellow type evolved from the purple one as a natural mutant. The purple cultivars grow better at higher altitudes in the tropics, while the yellow types grow better in lower areas. Passionfruit is cultivated in Argentina, Australia, Brazil, Mexico, New Zealand, South Africa, the United States, and the West Indies.

### Morphology

Passionfruit belongs to family Passifloraceae and has diploid chromosome number 2n = 18. It is a woody, vigorous, dicotyledonous, perennial climber growing up to 15 m long. Leaves are dark green, palmate, and three lobed with axillary tendrils. Flowers are perfect, single, axillary, and fragrant with showy calyx; they are tubular in shape, with five petals and five stamens, and the ovary has single locule. Fruit is a berry, globose or ovoid, deep purple with many dots, hard pericarp. Seeds are numerous, flat, and surrounded by yellowish pulp and a juicy aril.

## SEED STORAGE

Seeds are largely used for propagation of the plant. They are also used in evolving new cultivars through hybridization and in the conservation of genetic diversity. Seeds are also eaten as a food along with the pulp. Passion fruit seeds show orthodox storage behavior and tolerate loss of moisture and chilling during storage. However, seeds deteriorate rapidly under high humidity tropical conditions.

### Seed Collections

Passion fruit grows well in hot humid climates and in a wide range of soils. Seeds are sown in a nursery. Seedlings are transplanted to the field after three to four months of sowing. Suitable supports, such as trellises or poles, are provided for optimum vine growth (see Figure 16.1). Plants flower on new growth, and bees cause pollination. Fruit normally takes ten weeks for development. Healthy and ripe fruits are selected for seed purposes.-Seeds are separated from fruit, washed thoroughly, dried, cleaned, and packed for storage.

### Seed Germination

Seed germination is slow and non uniform. Fresh seeds show a considerable amount of dormancy due to the presence of a hard seed coat. Seed scarification using sand paper and removal of the hard seed covering promotes germination. Further, percentage of germination improves with exposure to alternate temperatures of 20/30°C for 16/8 h for six weeks. In normal conditions, seeds take 12 weeks for seedling emergence.

*Figure.1. Passionfruit Vines Grown on Trellises*

## Storage Conditions

Seed deteriorates rapidly at higher storage temperature and higher relative humidity. Costa, Oliveira, and Lellis reported that seeds stored at room temperature gave higher germination after 240 days of storage after drying seeds with pulp in sun or without pulp in shade. In open storage, seeds lose viability rapidly. Seeds with 5.2 percent mc stored at room temperature lost viability completely after ten months of storage.

Low-moisture seeds packed in polyethylene bags and stored in cool conditions preserved high viability for longer periods. Seeds stored in plastic bags in an air-conditioned room maintained high viability and exhibited 72 percent germination after 12 months of storage. Oliveira, Sadar, and Zampieri observed loss of viability to 50 percent after five years of ambient storage in paper bags, and none of the seeds germinated during the sixth year. Seeds packed in polyethylene bags and stored in cool chambers gave 60 percent germination after 57 months of storage, unlike none after 32 months in paper bags in ambient conditions.

# Chapter 17

# SUGAR APPLE: *ANNONA SQUAMOSA* L.

## INTRODUCTION

Sugar apple is also called custard apple or sweetsop. It is cultivated for the edible fruits. Ripe fruits are eaten fresh and used in the preparation of juice and ice cream. Sugar apple originated in South America and is now widely grown in Asia and Africa. It is presently cultivated in Brazil, Costa Rica, India, Indonesia, the Philippines, and Seychelles. A member of the family Annonaceae, it has chromosome number 2n = 14. It is a small tree growing up to 5 m in height. Flowers are perfect and yellowish in color. Fruit is a fleshy syncarp formed by the fusion of pistils and receptacle that is yellowish green, heart shaped, and covered with fleshy tubercles. Pulp is sweet, whitish, and embedded with numerous hard seeds. Pulp contains 16 to 18 percent sugar. Seeds are commonly used for propagation, and seedling trees are heterogeneous. Flowers are protogynous, and pollination is by insects. Ripe fruits are soft and perishable.

## SEED STORAGE

Sugar apple seeds are orthodox in storage behavior and preserve fairly well under ambient conditions. Seeds show a short period of dormancy. The embryo is small, partially developed, and embedded in a large endosperm. It continues to develop even after shedding from the mother plant, and one to three months are required for after-ripening of seeds. About 60 percent of seeds are viable after two months of storage. Low-moisture seeds packed in moisture-proof containers at low temperatures maintain viability for longer periods.

# Chapter 18

# SOURSOP: *ANNONA MURICATA* L.

## INTRODUCTION

Soursop is commonly cultivated for the edible fruits. Fruit is eaten fresh after peeling the exocarp. Flesh is used in preparation of juice. Soursop is native to South America and cultivated throughout the tropics. It is largely grown in Surinam, the United States, and Venezuela. Soursop is a small evergreen tree belonging to the family Annonaceae and has chromosome number 2n = 16. Flowers are yellow, protogynous, and strongly odorous.

Fruit is dark green, ovoid, and covered with curved fleshy spines. Pulp is soft white, with a sweetish sour taste, and contains many black seeds. It is propagated by seeds and grows well in sandy loam soils. It is cross-pollinated by insects.

## SEED STORAGE

Soursop seeds show orthodox storage behavior. Seed loses viability rapidly at higher temperature and relative humidity. Seed viability is lost completely after 210 days of storage at ambient temperature or 30°C, while it can be preserved for 390 days when stored in cloth, paper, plastic, or glass containers at 5 or 20°C.

# Chapter 19

# DATE PALM: *PHOENIX DACTYLIFERA* L.

## INTRODUCTION

Dates are a popular dessert fruit in North Africa. They are cultivated for dried date fruits and also grown as ornamental plants in the semiarid tropics.

Fully ripe dried dates are eaten fresh. Fruit is high in sugar, calcium, potassium, and iron contents. Plants require high heat during growing and ripening. Date palm originated in Central Asia, particularly in Iraq and surrounding areas. It is cultivated predominantly in the tropics and subtropics of western Asia and North Africa. Major date-producing countries are Algeria, Bahrain, Egypt, Iran, Iraq, Libya, Morocco, Pakistan, Saudi Arabia, Sudan, Tunisia, and the United States.

Date palm belongs to family Palmae. The diploid chromosome number is 2n = 36. Tree is slender, tall (30 m), dioecious, and erect with ascending foliage. Leaves are narrow and pinnate. Flowers are small and yellowish brown. Fruit is reddish brown, oblong or cylindrical, 2 to 8 cm long with sweet edible pulp and a single pointed grooved seed. Tree is propagated by offshoots or seedlings. It is wind pollinated, and fruiting occurs five to six years after planting.

## SEED STORAGE

### Seed Germination

Seed germination is slow and delayed, owing to the hard seed surface. Soaking in concentrated sulfuric acid for 30 min hastens seed germination.

### Storage Conditions

Date palm seeds survive for short periods under ambient conditions. Seeds exhibit orthodox storage behavior. High seed moisture and high storage temperature reduce the storage life of seeds. Seeds having 15 percent moisture stored in paper bags deteriorated after three months, and in poly ethylene bags at 3 to 4°C, seeds kept well for seven months. Fresh seeds show 91.8 percent germination after one month of storage, but this was reduced to 55.2 percent after nine months of storage. Nixon preserved date seeds for six years at an average temperature of 10°C in winter and 29.4°C in summer. Seed germination was 26 percent after 11 years, and after 15 years of storage, none of the seeds germinated.

# Chapter 20

# STRAWBERRY: *FRAGARIA X ANANASSA* DUCH.

## INTRODUCTION

Strawberry is a cool-season and cool-region crop. It grows well in cooler mountainous areas. It is largely cultivated in North America and Europe. The major strawberry-growing countries are Australia, Canada, England, France, Germany, Israel, Italy, Japan, New Zealand, Russia, Turkey, and the United States. Ripe fruits are eaten fresh or served with ice cream or in fruit salad.

Fruit is aggregate and rich in vitamin C and amino acids. Strawberry is a stoloniferous perennial herb belonging to family Rosaceae. The chromosome number is 2n = 16. Leaves are light green and large; flowers are yellow and are self- and cross-pollinated. Fruit is an aggregate of small achenes, sweet, and red to dark red. Seeds and runners are used for propagation. Seeds are extracted from ripe fruits, washed, dried, and packed for storage.

## SEED STORAGE

Strawberry seeds show orthodox storage behavior. Seed longevity improves with reduction in moisture and storage at lower temperatures.

## Seed Germination

Seed germination is slow and erratic, owing to dormancy. Fresh seeds of different varieties gave 9 to 99 percent germination. Dormancy varies in different cultivars. Seeds produced by self-pollination are much more dormant than seeds produced from cross-pollination. Seed germination is improved with dry storage for three months, but this did not eliminate the dormancy. Brown and Musa reported that seeds extracted from rotted fruits were less dormant, and seed drying at room temperature (18 to 20°C) promoted dormancy.

Precooling of seeds eliminates the dormancy. Seeds exposed to 4°C for four weeks and in combination with GA (100 ppm) showed higher germination. Light is essential for promotion of germination in dormant seeds. Red light promotes, and far-red light inhibits, the germination process opined that white light is sufficient for practical purposes in enhancing seedling emergence. According to Negi and Singh , inhibitors were of nonphenolic type and removed by continuous washing of seeds in water. Seed soaking in ethephon (5,000 ppm) for 24 h followed by chilling for 30 days resulted in higher seed germination. Iyer, Chacko, and Subramanyam reported that ethephon replaced the chilling requirement and gave higher germination. Further, seeds sown in intermittent mist gave early and uniform seedling emergence. Seed germination is also improved with soaking seeds in nitric acid and hydrogen peroxide. Seed scarification (sulfuric acid for 10 min) followed by chilling (4 to 6°C for five months) improved the germination percentage.

## Storage Conditions

Strawberry seeds are stored for two years under ambient conditions. Seed viability was higher in seeds stored in cooler locations. Seed viability is retained for 17 years in cool storage. Seed germination was early and rapid in older seeds. Low-temperature storage increases the storage life of seeds. Seeds preserved high viability and vigor at 4.4°C for two to three years.

# Chapter 21

# DURIAN: *DURIO ZIBETHINUS* L.

## INTRODUCTION

Durian is a fruit crop largely cultivated in Southeast Asia, especially in Malaysia. It is cultivated for the edible fruits. The sweet arils are eaten fresh or used in the preparation of jam or fermented products. Seeds are eaten after boiling and roasting. Pulp is rich in carbohydrates and fat, and it quickly turns sour and rancid.

Durian originated in Southeast Asia, particularly in the Malaysian peninsula. It is cultivated predominantly in Indonesia, Kampuchea, Malaysia, the Philippines, Thailand, and Vietnam.

### Morphology

Durian belongs to family Bombacaceae, and its chromosome number is 2n = 56. It is a large evergreen tree. Flowers are white or pink in color and arranged in cymes. Fruit is a large, spherical, ovoid, spiny, green to yellow capsule that splits into five valves. Seeds are large and covered with a sweet, pulpy, cream-colored, edible aril with one to six seeds per cell. Durian grows well in hot, humid climates with high rainfall. It prefers sandy loam soil. Seeds and vegetative means are used for propagation. It fruits seven years after planting. Moths and bats pollinate it. Fruit takes about three months to develop. It emits a bad odor similar to a rotten onion immediately after ripening.

## SEED STORAGE

Durian seeds remain viable for a short period under ambient conditions. Fresh seeds germinate readily within three days of sowing. Durian seeds exhibit recalcitrant storage behavior. They lose viability on drying and exposure to chilling temperatures. Fresh seeds contain about 44 percent moisture and lose viability when moisture falls below 21 percent. Further, a longer period of moist storage causes sprouting and rotting of seeds. Seed storage in airtight containers at 20°C preserves viability (90 percent) for 32 days.

# Chapter 22

# PHALSA: *GREWIA ASIATICA* MAST.

## INTRODUCTION

Phalsa is a delicate, easily perishable fruit. Ripe fruits are eaten fresh and used for juice preparation. Phalsa belongs to family Tiliaceae and originated in India. It is cultivated in India, Malaysia, Pakistan, and Taiwan. Fruit is somewhat acidic and contains high amounts of vitamins A and C and minerals such as calcium, phosphorus, and iron.

Phalsa is a bushy deciduous tree growing up to 10 m in height. Flowers are perfect; sepals number four to five; petals number four to five; stamens number 70 to 80; there is one pistil. Fruit is round, drupe, red or purple in color, and contains one to two seeds. Seeds are used for propagation. Fruits are harvested when color changes to reddish brown. Seeds are extracted from ripe fruits, washed, dried, and suitably packed for storage.

## SEED STORAGE

Seeds show orthodox storage behavior and maintain greater longevity under dry, cool conditions. Fresh seeds show dormancy, which terminates with after-ripening. Phalsa seeds are short-lived under ambient conditions. They lose viability rapidly at higher temperatures and humidity. Seed viability is unaffected by drying in sun or shade.

Verma and Rathore reported that seeds sown after four weeks of storage resulted in higher germination as compared to fresh seeds, whereas none of the seeds germinated after ten weeks of storage under ambient conditions. Seeds preserved higher viability when packed in sealed jars and stored at 5 to 8°C or room temperature than when stored in open conditions at the same temperature. Well-dried seeds (8 percent mc) packed in polyethylene bags stored at 5 and –20°C maintained viability (60 percent) for ten years.

# Chapter 23

# MACADAMIA NUT: *MACADAMIA INTEGRIFOLIA*

## INTRODUCTION

Macadamia is cultivated for the edible seed kernels, which are eaten as a dessert or used in ice creams, chocolates, and confectioneries. It is rich in fat. Macadamia is native to Australia, and largely cultivated in Australia, Brazil, Costa Rica, Guatemala, Kenya, South Africa, the United States, and Zimbabwe.

Macadamia is an evergreen tree belonging to family Proteaceae. Flowers are small, perfect, and borne in large numbers (100 to 300). Fruit is a follicle, seeds are spherical with hard testae, and the kernel is creamy white and rich in fat. Macadamia grows well in high rainfall areas in the tropics. Many cultivars are self-incompatible and pollinated by honeybees. Fruits that fall on maturity are collected for storage.

## SEED STORAGE

Seeds are used for propagation as well as for raising rootstocks. Macadamia seeds appear to be recalcitrant and viable for a short period under ambient conditions. Seed viability varies from 42 to 83 percent in different cultivars. Fairly longer periods (40 to 240 days) are required for seedling emergence following sowing. Seeds are viable for four months at room temperature. Hamilton reported that seed viability was preserved for four months at room temperatures (6.7 to18.3°C), but it subsequently deteriorated and was lost completely by 12 months of storage. Similarly, moist seeds retained 50 percent of germination after six months of storage under ambient conditions.

# Chapter 24

# INDIAN JUJUBE: *ZIZIPHUS MAURITIANA* LAM.

## INTRODUCTION

Indian jujube is also known as ber. It is grown mainly for the edible fruits. The fruit is eaten fresh, dried, or canned. It is rich in minerals, such as iron, phosphorus, calcium, and in vitamins A and C. Leaves, bark, and roots are used for medicinal purposes. The wood is a valuable timber in making furniture. Indian jujube originated in India. Later it spread to Africa and Southeast Asia. It is commonly grown in China, India, Indonesia, Malaysia, Mozambique, Pakistan, the Philippines, Sri Lanka, and the West Indies.

### Morphology

Indian jujube belongs to family Rhamnaceae and has chromosome number 2n = 48. It is a small, thorny evergreen tree. Leaves are simple, alternate, and broadly oval. Flowers are small, greenish, hermaphrodite, and borne in clusters at the leaf axils. Fruit is ellipsoid to subglobose drupe, fleshy, and orange to brown in color with edible acidic pulp and a hard central stone. The hard endocarp encloses one or two seeds.

## SEED STORAGE

Seeds are predominantly used for propagation. They show orthodox storage behavior and maintain viability for fairly long periods under ambient conditions. However, unfavorable storage conditions reduce seed longevity. Seeds are also used for raising rootstocks for vegetative propagation by budding or grafting and in the conservation of genetic diversity.

## Seed Collections

The Indian jujube grows well in hot, dry climates. High humidity and frost damage the crop, especially during the fruiting stage. It prefers deep, sandy loam soils. However, it withstands waterlogging and drought conditions. Plants are raised through seeds or patch budding. Honeybees cause pollination. Fruit takes four to five months to mature. Large healthy ripe fruits are used for seed purposes. Seeds are extracted, washed, dried, and suitably packed for storage.

## Seed Germination

Seed germination is slow and uneven, owing to the hard endocarp. The seed is covered by the endocarp, which hardens after 50 days of bloom, and the testa turns to dark brown 100 to 110 days after anthesis. Moisture content decreases to 30 percent at harvest. Immature seeds do not germinate, while treatment with benzyladenine or gibberellic acid (10 ppm) stimulates germination. Seed germination is low in fresh seeds but improves on soaking in water or in acid. Seed soaking in concentrated sulfuric acid for six minutes gave highest seedling emergence. Casini and Salvadori reported that removal of the endocarp and treatment with gibberellic acid (400 ppm) improved germination, and seed with the endocarp intact germinated after scarification by immersion in concentrated sulfuric acid.

## Storage Conditions

Fresh seeds gave low germination (33 percent), even up to eight months of storage. Seedling vigor was high at the end of the storage period. Seed viability and vigor are retained with low-temperature storage. About 50 to 60 percent of seeds were viable after ten years of storage at 5 and –20°C. Seed viability improved in one-year-old seeds by soaking them in water for 48 h followed by four days' stratification in moist sand. Seed treatment with potassium dihydrogen phosphate (1 percent) improved germination in stored seeds (203 days after extraction).

# Chapter 25

# MULBERRY (*MORUS* SPP.)

## INTRODUCTION

Mulberry is cultivated for the edible fruits in many parts of the tropics and subtropics. It is cultivated in Canada, China, India, Japan, and the United States. It belongs to family Moraceae. The different species include *Morus alba, M. indica, M. latifolia, M. lhou, M. nigra,* and *M. rubra.* The black *M. nigra* is well suited for higher altitudes in the tropics. Ripe fruits are eaten fresh as a dessert and also cooked as a vegetable. Tree is monoecious or dioecious, small, and with thin, ovate, serrate leaves that are often deeply lobed. Fruit bunch is syncarpous with many drupes, 2.5 to 5.0 cm long, and white, red, or pink in color. Seeds or cuttings are used for plant propagation.

## SEED STORAGE

Seedling emergence is slow and nonuniform due to the presence of dormancy. Two years of dry storage over calcium chloride (CaCl2) removes the dormancy, especially in *Morus alba.* Seed exposure to alternate temperatures of 20/30°C for 16/8 h promotes germination. Further, seed irradiation with gamma rays (7.5 kilorad [kr]) improves the germination.

Mulberry seeds show orthodox storage behavior and are short-lived under ambient conditions. High seed moisture is injurious to seed longevity, and viability is lost rapidly at room temperature. Seeds stored in desiccators over CaCl2 + 20 percent H2O (water) or 43.8 percent H2SO4 (sulfuric acid) remained viable for 14 years as compared to 4 months in control. The relative humidity in desiccators was 48 percent, and seed moisture content was 4 percent. Low-moisture seeds packed in moisture-proof containers at low temperatures retain viability for fairly long periods.

# Chapter 26

# SAPODILLA: *MANILKARA ACHRAS* (MILL.) *FOSBERG*

## INTRODUCTION

Sapodilla is an important tropical fruit cultivated for its edible sweet pulp. Normally, the ripe fruits, which are rich in sugars, calcium, phosphorus, and iron, are eaten fresh. Unripe fruits are sources for extraction of chicle gum. Sapodilla belongs to family Sapotaceae and is native to Mexico. It is cultivated in Brazil, Cuba, India, Malaysia, Mexico, the Philippines, Sri Lanka, Surinam, Thailand, the United States, Venezuela, and the West Indies. Sapodilla is a hardy evergreen tree requiring less attention for its cultivation.

It grows well in hot humid climates, especially in tropical lowlands. Fruit is globose to ovoid and rusty brown; the yellowish brown flesh contains nil to 12 black seeds. Seeds are hard and easily separable from the pulp.

Seeds, layers, and grafts are used for raising sapodilla plants. Sapodilla seedlings grow slowly; hence, it is replaced with *Manilkara hexandra* as a rootstock.

## SEED STORAGE

Sapodilla seeds have limited use. They are used for raising rootstocks, and in hybridization programs. Seeds are viable for a short period. It appears that sapodilla seeds are recalcitrant in storage behavior. They lose viability rapidly on removal from fruit. Seedling emergence is rather slow and uneven, possibly due to the hard seed coat. Seed germination varies in different cultivars. It was highest in cv. Periyakulam (PKM) (58 percent) and least in cv. Dwarapudi (26 percent). Seed germination was improved by application of gibberellic acid (50 ppm), indoleacetic acid (50 ppm), and thiourea (1 percent).

# Chapter 27

# RAMBUTAN: *NEPHELIUM LAPPACEUM* L.

## INTRODUCTION

Rambutan is largely grown in Southeast Asia. Fruits have a sweet, acidic, edible aril, which is eaten fresh, canned, or mixed with other fruits. Seeds, which are eaten after roasting, are rich in fat and are hence used in manufacturing soap and candles. Different plant parts are used for medicinal purposes.

Rambutan is native to Malaysia. It is cultivated in Australia, Honduras, Indonesia, Malaysia, Mexico, the Philippines, and Thailand. Rambutan belongs to family Sapindaceae and has chromosome number 2n = 22. It grows well in hot, humid tropical conditions with high rainfall. Its cultivation is successful in its native area. It prefers well-drained soil. Trees are evergreen, monoecious or dioecious, and grow up to 20 m in height. Flowers are unisexual or perfect, small, greenish white, and arranged in panicles. Fruit is globose or ovoid with hairy prickles. Its flesh is juicy, white, sweet, or sourish. Seeds are globose and 2.5 to 3.5 cm in diameter.

Rambutan is propagated by seeds and by bud grafting. Seedling trees take longer to bear fruit, five to six years, and are not to true to type. Insects such as bees and flies pollinate the flowers. Fruits are nonclimatric and harvested when fully ripe. Seeds are removed from fruit, washed, and used immediately for propagation or for short-term storage.

## SEED STORAGE

Rambutan seeds are used for propagation as well as for raising rootstocks. Seeds are recalcitrant and do not store well in ambient conditions. Fresh seeds contain a high amount of moisture (32.5 percent) and germinate readily (100 percent). Seedling emergence was rapid when placed horizontally. Seeds lose viability within one week of storage under ambient conditions due to drying. During ambient storage the moisture decreases to 13 percent after two weeks and seeds are killed.

Seed germination is affected in the presence of a fruit arilloid. The juice of the arilloid also inhibits germination. However, seeds are preserved well in moist sawdust or charcoal for three to four weeks. They sprout readily during storage. Unwashed seeds or seeds treated with juice from the arilloid remained viable for four weeks in moist sawdust without sprouting.

# TEMPERATE FRUITS

# Chapter 28

# APPLE

## INTRODUCTION

Apple is the leading temperate fruit crop of the world. It is primarily cultivated for the edible fruits. Fruits are eaten fresh or processed as jam, juice, and cider. Fruits are rich in sugar and also contain high amounts of minerals, such as calcium, phosphorus, iron, and potassium, and vitamin thiamine. Trees have showy flowers and are used in landscaping and in home gardens for ornamental purposes.

### Origin and Distribution

Apple is largely cultivated in Europe and North America. It is likely derived from *Malus pumila* in southwestern Asia. The major apple-growing countries are Australia, Brazil, Canada, China, Denmark, Egypt, England, France, India, Israel, Italy, Japan, Mexico, New Zealand, Russia, and the United States.

### Morphology

Apple belongs to family Rosaceae, and its somatic chromosome number is 2n = 34. Trees are of medium size; leaves are soft and less pubescent; flowers are small, borne in clusters, and bright pink in color. Fruit is large and sweet and contains brown seeds. A thousand seeds weigh 25 to 50 g.

## SEED STORAGE

Apple is propagated by both sexual and asexual methods. Seeds are predominantly used for raising rootstocks but are also involved in evolving better varieties through hybridization. Seeds survive for fairly long periods under ambient conditions. They exhibit orthodox storage behavior. Fresh seeds do not germinate readily due to innate dormancy, and they require cold stratification for removing dormancy. Dormancy is helpful in extending storage life of seeds for short periods. Normally, apple fruits are stored for several months in cold storage, which helps to preserve seed quality.

## Seed Collections

Apple is a cool-season crop and is widely grown in cooler regions, particularly at higher latitudes and altitudes. It requires chilling temperatures for induction of flowering. It prefers deep, fertile, and well-drained soil. Apple is commonly propagated by grafting using suitable dwarfing rootstocks.

Many apple cultivars are self-incompatible and need suitable pollenizers. Honeybees largely bring about pollination. The grafted plant bears fruit in two to four years after planting. Ripe, healthy, good-sized fruits are collected for seed purposes. Seeds are removed from the fruit, thoroughly washed, dried at low temperature and relative humidity, and suitably packed for storage.

*Seed storage in fruit.* Apple seeds preserve viability and vigor in fruits up to 180 days. Fruits stored for shorter periods showed high seed quality. The longer the seeds remained in the fruit (150 to180 days), the higher their susceptibility to mildew. Seeds stored in fruit for 30 days showed maximum seedling growth and development. Seedling emergence in seeds extracted from fruits kept in cold storage was higher than for those kept at room temperature. It is advantageous to preserve fruits at 5°C for a short period prior to seed extraction.

## Seed Germination

Apple seeds exhibit dormancy and do not germinate immediately after extraction. Storing fruits at low temperature eliminates the dormancy, but it increases time to germination. Seed dormancy can be overcome by exposing moist seeds to low temperature (5°C) for various periods. According to Abbot , germination of dormant seeds is higher at lower temperatures (5 to 10°C). Radicles emerge at lower temperatures and hypocotyls develop at higher temperatures. Seeds of late-harvested fruit and rotten fruits need shorter periods of stratification.

Seed coat removal facilitates germination but produces more abnormal seedlings. Normal growth is resumed after a certain period or through stratification of seeds or seedling exposure to low temperatures. The longer period of stratification is reduced by growing excised embryo or treating seed with ethephon (250 ppm) followed by 60 days exposure to low temperature, which gave high seed germination.

Early and rapid techniques for determination of viability in apple seeds have been developed. Seeds are soaked for 2 to 3 h in sulfuric acid, followed by soaking in benzyladenine (10 ppm) or equal volumes of benzyladenine (15 ppm) and gibberellic acid (150 ppm). Seeds germinate in the presence of continuous light at room temperature. The viable embryo germinates in three to four days of sowing, while nonviable ones deteriorate.

## Storage Conditions

*Temperature.* Apple seed exhibits orthodox storage behavior. Higher seed moisture and temperature reduce the storage life of seeds. The estimated storage life of apple seeds with 5 percent mc is 100 years and 37 years at −18 and 5°C, respectively. Seed moisture is lowered rapidly in apple seeds by exposing them to low relative humidity. It takes about eight days to reduce moisture from 50 (fresh seeds) to 5 percent. Seed germination decreases with increase in moisture during storage. Seeds stored in a damp atmosphere show low germination, while seeds stored at 50 to 55 percent RH exhibit higher germination.

Grzeskowiak, Miara, and Suszka recommended sealed storage of apple seeds at −3°C for longer retention of viability and vigor. According to Kaminski , drying stratified seeds at a higher temperature (25°C) induces secondary dormancy, not the lower temperature (5°C). Therefore, Conner recommends storing seeds in moist sand at 0°C and not allowing the seeds to dry. Cromarty, Ellis, and Roberts suggested drying seeds at 15°C with 10 percent RH for storage. Seeds (8 to 10 percent mc) were stored at 10°C in jute or polyethylene bags up to 25 months. In jute bags, viability was reduced from 90 to 30 percent within nine months, and in polyethylene bags, seeds showed 90 to 95 percent germination after 25 months of storage. Seed viability can be preserved for two years in desiccators with calcium chloride at 2 to 10°C. Seed longevity is extended further to seven years by packaging seeds in hermetically sealed bottles and storing them at −3 to −5°C. Seed drying to 5 percent moisture and storage at −18°C are beneficial for longer retention of high seed viability and vigor.

## Invigoration of Stored Seeds

Seedling emergence is slow and less with an increase in storage period. Seed stratification in moist sand at 5°C followed by seed soaking in gibberellic acid (100 milligrams per liter [mg·liter −1]) for 12 h increased the seed viability in three-year-old apple seeds.

# Chapter 29

# PEAR

## INTRODUCTION

Pear ranks second after apple as a popular cultivated fruit in temperate regions. The two important cultivated are *Pyrus* species *P. communis* and *P. pyrifolia.* Pear cultivation is also extended to subtropical regions due to its hardy nature and wider adaption. Fruits are used for table purposes and processed for preparation of jam, jelly, juice, wine, and candied and sweet pickles. Fruits are rich in organic acids, such as malic and citric acids; minerals, such as copper, sulfur, and phosphorus; and vitamins, such as thiamine, riboflavin, and vitamin C. Pear flowers are beautiful and the trees are planted in house gardens for ornamental purposes.

### Origin and Distribution

Wide genetic diversity exists among pear cultivars in Europe and Asia. Pear originated in eastern Asia, probably in western China. European cultivars are descendants of *P. communis,* while Chinese and Japanese cultivars are derived from *P. pyrifolia.* Pear cultivation is mainly confined to temperate and subtropical regions of the northern hemisphere. It is grown in Argentina, Australia, China, France, Germany, Italy, Korea, Romania, Russia, South Africa, Spain, Switzerland, Turkey, and the United States.

### Morphology

Pear belongs to family Rosaceae, and its somatic chromosome number is 2n = 34. The commonly cultivated varieties belong to *P. communis* and *P. pyrifolia,* while *P. serotina* and *P. pashia* are used as rootstocks. Pear is a small deciduous tree. Leaves are small, ovate, and margin serrated; flowers are showy, hermaphrodite, and white. Fruit is a pome, is pyriform and fleshy, and contains a stone. Seeds are formed inside the stones.

## SEED STORAGE

Pear is commonly propagated by vegetative means such as budding and grafting. Seeds are primarily used for raising rootstocks. Seeds are also used in evolving better cultivars through hybridization and for conservation of genetic diversity in the gene bank. Pear seeds exhibit orthodox storage behavior, and their longevity is extended by lowering seed moisture and storing at low temperatures.

### Seed Collections

Pear plants require mild winters and summers for optimum growth. They grow well in deep, well-drained, sandy loam soil. Trees bear fruit five to six years after planting. Many pear cultivars are self-incompatible and need suitable pollenizers. Bees bring about pollination. Fully matured, healthy fruits are collected for seed purposes. Seeds are easily extracted after softening the fruit. They are washed thoroughly, dried, cleaned, and suitably packed for storage or stratified for immediate sowing to raise seedlings for rootstocks.

### Seed Germination

Fresh pear seeds do not germinate readily and need longer stratification to stimulate the germination process. Seed dormancy is removed by storing moist seeds at low temperature. Seeds stored at 5°C for 150 days showed high germination in many pear cultivars. The inhibitor contents decrease and auxinlike substances increase in seed coat and embryo during stratification. The duration of stratification varies from 15 to 180 days in different species. The stratification period also depends on the location from which fruits are collected.

Seeds from warm winter climates require less chilling than those from colder climates. Seed germination is high after removal of the seed coat because this causes the excised embryo to germinate quickly. The seedlings initially show abnormal growth if seeds have not been stratified. Shen and Mullins reported that moist seeds placed at 4°C for four weeks, and subsequently at the soil surface in the presence of dim light, gave higher germination. Seed germination is also improved with seed soaking in gibberellic acid (150 ppm) and thiourea (5,000 ppm), followed by 28 days of stratification.

### Storage Conditions

Pear seeds readily absorb moisture from the atmosphere. High seed moisture reduces storage life of seeds. It takes eight days for seed moisture to decrease from 50 (fresh seeds) to 5 percent. Low moisture (5 to 6 percent) is ideal for seed storage. The drying of stratified seeds results in low germination due to the induction of secondary dormancy. Seeds germinate fully when stratified. Pear seeds are dried at low temperature (15°C) and low relative humidity (10 percent RH) for long-term storage. Seed viability is preserved for two years when seeds are packed in sealed bottles or stored in desiccators over calcium chloride at 2 to 10°C. Seed longevity was extended further to seven years by preserving seeds at –3 to –5°C. Seeds preserved in liquid nitrogen (–196°C) maintain viability for three years. Seed viability was successfully preserved for 22 years at –5°C. For better and longer storage of pear seeds, Ellis suggested storing fruits initially at 5°C for a short period prior to seed extraction, then reducing moisture content to 5 percent before packing in suitable moisture-proof containers and storing at subzero temperature (–18°C).

# Chapter 30

# PEACH

## INTRODUCTION

Peach is a delicious fruit grown in temperate regions. The edible juicy fruits are eaten fresh and processed for squash preparations. The short stature of the plant and its showy flowers are valued for ornamental purposes, and it is thus planted in home gardens for landscaping. The wide genetic diversity of peach, including wild types, is found in China, from where it is assumed to have originated. Peach is predominantly cultivated in Argentina, Australia, Chile, China, England, France, Japan, Mexico, New Zealand, South Africa, Spain, Turkey, and the United States.

### Morphology

Peach belongs to family Rosaceae and has diploid chromosome number 2n = 16. Trees are small; leaves are long, simple, deciduous, and alternate. Flowers are perfect and pink to red in color. Fruit is a spherical, fleshy drupe with juicy outer layer surrounding hard stones that contain seeds. Trees are short-lived, lasting eight to ten years, and they commence fruiting from the third year onward.

Budding is followed for propagation of peach plants, and seeds are used for raising rootstocks. Peach is a cross-pollinated crop, and insects bring about pollination. Fruit changes color on maturity. Seeds are separated from matured fruits, washed thoroughly in water, dried, and suitably packed for storage.

# SEED STORAGE

## Seed Germination

Seed germination and seedling emergence are higher in mature seeds. Deol, Chopra, and Grewal reported that seeds collected from fruits harvested 10 to 12 days before commercial maturity had significantly higher germination than the seeds of fruits collected at commercial maturity.

Peach seeds show dormancy and need suitable dormancy-breaking treatments for higher germination. The hard endocarp delays seedling emergence, and removal of the endocarp results in early germination. Prechilling of seeds is beneficial for promoting seed germination. Exposure of moist seeds to 3 to 5°C for three to four months eliminates dormancy. Warm stratification for a short period followed by cold stratification promotes germination in dormant seeds. Suszka reported that preserving the seeds for 14 days at 20°C followed by prechilling at 3°C resulted in higher germination.

## Storage Conditions

Peach seeds show orthodox storage behavior. Reducing seed moisture and storage at low temperatures help maintain seed viability. Cromarty, Ellis, and Roberts recommended that seed moisture could be effectively reduced at a low temperature (15°C) and seed stored at low relative humidity (10 percent RH) without damaging seed quality. Seeds are successfully dried to 8 percent mc. Seeds with intact endocarp take longer to dry; hence, the endocarp is halved carefully before drying. Earlier, seeds were stored under moist conditions for short storage periods. Seeds with 8 to 12 percent mc were preserved at 3°C for more than eight weeks. Ellis and Hong recommended storage of dry seeds in moisture-proof containers at -20°C for longer storage life of seeds.

# Chapter 31

# PLUM

## INTRODUCTION

Plum is a delicious juicy fruit grown in temperate regions. Fruits are used for table purposes, canned, dried, and processed for jam, jelly, squash, and confectionery. Plums are a rich source of sugar and contain high quantities of vitamins A, thiamine, and riboflavin and minerals such as calcium, iron, and phosphorus. Trees are planted in home gardens for ornamental purposes.

The different types of cultivated plums include American plum *(Pru nus americana)*, cherry plum *(P. cerasifera)*, damson plum *(P. insititia)*, European plum *(P. domestica)*, and Japanese plum *(P. salicina)*. Plum originated in North America, China, and Europe. The major plum-growing countries are Austria, Bulgaria, China, France, Germany, Hungary, Italy, Japan, Mexico, Romania, Russia, Spain, Turkey, and the United States.

### Morphology

Plum belongs to family Rosaceae and has somatic chromosome number 2n = 16. Tree is small; leaves are ovate; flowers are white; fruit is oblong to ovoid and variously colored. Most plum cultivars are self-incompatible and require pollenizers. Bees bring about pollination. Ripe, healthy, and firm fruits are selected for seed purposes. Seeds are removed from pulp, washed thoroughly with water, dried, and packed in suitable containers for storage. Propagation is commonly achieved by budding; seeds are used in raising rootstocks.

# SEED STORAGE

## Seed Germination

Seed germination is slow and it takes longer for seedlings to emerge. Fresh seeds show dormancy. Germination is improved by removal of the seed coat. Prechilling of imbibed seeds also helps to break seed dormancy. Seed exposure to high temperature (20°C) for 14 days followed by chilling at 3 to 5°C hastens the germination process.

## Storage Conditions

Plum seeds show orthodox storage behavior. Dry seeds preserve high viability for long periods at cooler conditions. Giersbach and Crocker  preserved seeds under moist conditions at 7 to 10°C. Seed germination was 98 percent after 26 months and 16 percent after 53 months of storage. Ellis and Hong  recommended seed drying at 15°C and 10 percent RH to 4 to 6 percent mc. Dried seeds are packed in moisture-proof containers and stored at –20°C for long-term seed conservation.

# Chapter 32

# CHERRY

## INTRODUCTION

Cherry is commonly cultivated in temperate regions. Primarily two types of cherries are grown, namely, sweet cherry *(Prunus avium)* and sour cherry *(P. cerasus).* Sweet cherry is mainly used for table purposes, but also in sal ads and juice preparations and for canning. Sour cherry is largely used for processing purposes, such as wine or juice making, canning, and freezing; it is also sun dried. Fruit is rich in sugar, calcium, potassium, magnesium, iron, zinc, and vitamins A and C.

### Morphology

Cherry has its origin in southeast Europe. It belongs to family Rosaceae, and its diploid chromosome number is 2n = 16, 32. Trees are tall; flowers are white and perfect. Fruits are globular or oblong and red, with a fleshy outer layer surrounding the hard stones that contain the seeds. Cherry is propagated by budding and grafting. Seeds are used for raising rootstocks.

Many cherry cultivars are self-incompatible and require pollenizers. The compatible cultivars are grown side by side. Honeybees bring about pollination. Fruits are harvested when they change color. Seeds are extracted from fully ripe fruits, washed thoroughly with water, dried, and packed for storage.

## SEED STORAGE

### Seed Germination

Fresh cherry seeds do not germinate due to the presence of innate dormancy. Seedling emergence is delayed due to the presence of a hard endocarp, and it improves with removal of the endocarp. Seed exposure to prechilling at 3 to 5°C hastens the germination process. The percentage of germination was higher at alternate temperature regimes of 3/5°C or 5/20°C for 16 and 8 h, respectively, than at a constant temperature of 3 or 5°C. Warm stratification before chilling is beneficial for promoting higher seed germination. Exposure of cherry seeds 14 days at 20°C prior to prechilling at 3°C promotes germination. Repetitions of the warm and cold stratification cycle result in full germination of seeds.

### Storage Conditions

Cherry seeds are orthodox and remain viable for longer periods at lowmoisture and low-temperature conditions. Seeds with 12 percent mc stored at 3 to 5°C preserved viability for five years. Seeds stored in hermetically sealed bottles at –3 to –5°C maintained 91 to 97 percent germination for seven years. Seeds lost viability rapidly when stored in sealed bottles at 20 to 25°C. Subsequently, high seed viability was preserved for 15 years at –5°C. Storage of dry seeds (4 to 6 percent mc) in moisture-proof containers at subzero temperature (-20°C) preserved viability for a longer period.

# Chapter 33

# RASPBERRY (*RUBUS* SPP.)

## INTRODUCTION

Raspberry is a hardy plant widely cultivated in temperate areas of the northern hemisphere. Red *(Rubus idaeus)*, black *(R. occidentalis)*, and purple raspberry (hybrid of red and black) are cultivated in different locations. Fruits are eaten fresh, canned, and used in the preparation of jam, jelly, syrup, and wine. Leaves are used for medicinal purposes. Raspberry is native to North America, Asia, and Europe. It is predominantly cultivated in England, Hungary, Russia, the United States, and Yugoslavia.

### Morphology

Raspberry belongs to family Rosaceae, and its diploid chromosome number is 2n = 14. Plant is erect and about 1 to 2 m in height. Suckers are commonly used for its propagation. Seeds are used for evolving new cultivars through hybridization. Flowers are small, few, and white. Fruit is oblong, conical, and dark red. Ripe and healthy fruits are selected for seed purposes. Seeds are extracted from the fruit, washed thoroughly with water, dried, and packed for storage.

## SEED STORAGE

### Seed Germination

Fresh raspberry seeds show poor and delayed germination, attributed to the presence of a hard seed coat. Seed scarification and removal of the seed coat hasten the germination process. Jennings and Tulloch reported that raspberry seeds germinated fully when scarified seeds were exposed to prechilling, alternate temperatures, or light or treated with potassium nitrate, thiourea, or kinetin. Further, gibberellins, sodium hypochlorite, calcium hypochlorite, and prolonged warm stratification promote the germination of intact seeds. Seed treatment with potassium hydroxide, sulfuric acid, or gibberellic acid stimulates germination.

Seed germination was higher with potassium hydroxide, and seeds soaked in concentrated sulfuric acid for more than 20 min were damaged.

## Storage Conditions

Raspberry seeds show orthodox storage behavior. Low seed moisture and low storage temperatures are beneficial for extending storage life of seeds. Seeds stored in polyethylene bags at 1°C remained viable for three months.

# Chapter 34

# BLUEBERRY (*VACCINIUM* SPP.)

## INTRODUCTION

Blueberry is an important small fruit in North America and Europe. It is cultivated for the edible fruits. Fruits are consumed fresh as well as processed. Products such as jam, jelly, sauce, pie, and wine are prepared. Plants are also used for ornamental purposes in home gardens. Fruits are a rich source of sugar, citric acid, vitamin C, phosphorus, calcium, iron, and potassium. Blueberry is native to North America. It is cultivated predominantly in Bulgaria, Canada, Chile, Denmark, Finland, Japan, the Netherlands, New Zealand, Poland, Russia, and the United States.

### Morphology

Blueberry belongs to family Ericaceae and has chromosome number 2n = 24. The commonly cultivated blueberries are highbush *(Vaccinium corymbosum)*, lowbush *(V. angustifolium)*, and rabbiteye *(V. ashei)* blueberry. The plant is a deciduous shrub, growing up to 6 m in height. Flowers are small, borne in clusters, and white or tinged with pink. Fruit is blue and sweet. Seeds, grafting, and cuttings are used to propagate blueberry bushes.

Lowbush and rabbiteye blueberries are self-incompatible and require crosspollination for higher fruit set. These are mainly pollinated by honeybees. Fruit becomes blue on full ripening. Fruit with larger seeds develops faster than fruit with smaller seeds. Healthy and fully ripe fruits are collected for seed purposes. Seeds are removed, washed thoroughly with water, dried in shade, and packed in moisture-proof containers for storage.

# SEED STORAGE

## Seed Germination

Blueberry seeds germinate readily on immediate extraction from ripe berries. Seed stratification at 3 to 4°C for 30 days promotes seed germina tion. Seed germination was higher when berries were stored at –23°C and seeds were stored in berries at –2°C or 1°C. The berries become moldy with *Botrytis* spp. and germination percentage is reduced considerably. Mature and large seeds give higher germination than immature and small seeds. Seed soaking in gibberellic acid (100 to 500 ppm) stimulates the germination process. Light also hastens the germination process in fresh seeds.

## Storage Conditions

Blueberry seeds preserve well under ambient conditions. High seed viability is preserved for two years at 5 to 7°C. Further, seeds preserved at 4.4°C gave higher germination (86.2 percent) after 12 years of storage.

# Chapter 35

# WALNUT (*JUGLANS* SPP.)

## INTRODUCTION

Walnut is native to Iran and the surrounding areas. It is extensively cultivated in the temperate regions of Bulgaria, China, France, Iran, Italy, Mexico, Romania, Turkey, and the United States, where summers are relatively warm. All plant parts of walnut are used. Immature fruits are used in the preparation of pickles, marmalades, and juice. Walnut is rich in proteins, fats, and minerals, such as phosphorus, potassium, and magnesium, and in vitamin B. Walnut oil is used in cosmetics. Trees are planted for landscaping and ornamental purposes. Its wood is highly valued for making various types of furniture.

### Morphology

Walnut belongs to family Juglandaceae and has somatic chromosome number 2n = 32. The various types of cultivated walnut are Persian walnut (*Juglans regia),* black walnut (*J. nigra),* and Heartseed walnut (*J. ailanthifolia).* are large, deciduous, monoecious, aromatic, ornamental, and long-lived. They grow up to 20 to 25 m in height. Nuts are enclosed within thick indehiscent husks. Walnut is pollinated by the wind. It is commonly propagated by seedling and vegetative means such as grafting and budding, and stooling is also employed. Seedlings are largely raised for rootstock purposes. It takes eight to ten years for fruiting. Hulls develop a crack on maturity. Healthy and mature nuts are collected for seed storage.

# SEED STORAGE

## Seed Germination

Walnut seed germination is slow and takes longer to complete. Seeds are dormant and need stratification to break the dormancy. Removal of a small part of the hard shell promotes radicle emergence.

Exposure of seed to longer chilling periods overcomes dormancy. Seed immersion in hot water (80 to 90°C) for 24 h followed by stratification for a short period results in uniform germination. Seed treatment with ethephon (100 ppm) improves the germination percentage.

## Storage Conditions

Walnut seeds show orthodox storage behavior. Seed longevity increases when seeds are stored under dry, cool storage conditions. Seed viability is higher (37 percent) in seeds stored loose in pots after four years of storage. In damp conditions (50 percent mc and 2.8°C) germination is less than 20 percent, but at lower seed moisture (30 percent) germination is as high as 76 percent.

# SEED STORAGE IN VEGETABLE CROPS

# Chapter 36

# POTATO

## INTRODUCTION

Potato is a vegetable crop with global importance. It has been acknowledged as a staple food crop in many countries and ranks fourth after rice, wheat, and maize in production. It is cultivated in temperate, subtropical, and tropical regions for its underground edible tubers. Potato is a good source of carbohydrates and proteins and also contains high quantities of potassium and vitamins such as niacin, thiamine, and riboflavin. It has industrial importance in the manufacturing of starch, and also in the making of potato chips.

### Origin and Distribution

Potato is native to South America. Its genetic diversity in terms of landraces and wild species occurs in Bolivia, Colombia, Ecuador, Peru, and Venezuela. Potato cultivation is distributed between 40°N and 20°S latitudes. It grows well in temperate regions, especially in Europe and North America. In the tropics, it does better at higher elevations. Besides North America, it is cultivated in Belarus, China, France, Germany, India, Japan, Kenya, Malaysia, the Netherlands, the Philippines, Poland, Russia, Sudan, Ukraine and United Kingdom.

### Morphology

Potato is an important member of the Solanaceae family. Plants are a dicotyledonous, perennial herb that is cultivated as an annual; the food is stored in tubers at the end of the stolons. Tubers vary in size, shape, color, and depth of eyes, which are characteristics of the different cultivars. They are bulky, perishable, and difficult to transport and store for long periods.

Potato stems are weak; leaves are alternate and pinnately compound with ovate leaflets. The cultivated potato is tetraploid (2n = 48) and diploid (2n = 24), while hexaploid (2n = 72) potatoes are also found. Flowers are small and white, yellow, blue, or purple and are borne in clusters. Fruit is a berry, globose, green or purplish green, and not edible (see Figure.1). It contains more than 300 seeds. Seeds are small (1.5 mm), yellow or brown, flat or kidney shaped, and poisonous. One thousand seeds weigh roughly 0.6 g.

## SEED STORAGE

Potato is propagated commercially by tubers, normally called seed tubers, to maintain the identity of the varietal characteristics, whereas true seeds are used in crop improvement and genetic conservation. Of late, cultivation through true potato seeds (TPS) is becoming popular in many potatogrowing areas, due to nonavailability of sufficient quantity of quality seed tubers, high transport costs, and problems in tuber storage. Potato seeds show orthodox storage behavior. These tiny seeds can be dried to a low level of moisture and stored at low temperatures for fairly long periods without affecting seed quality.

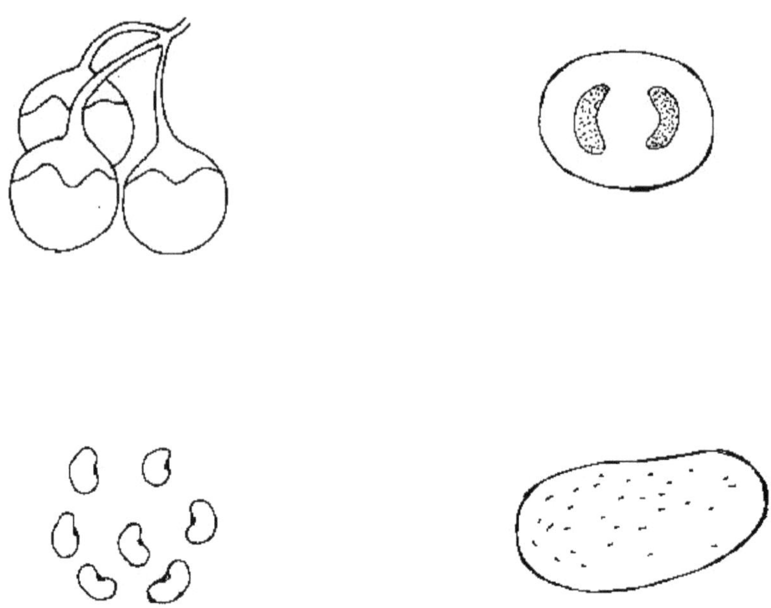

*Figure.1. True Potato Seeds and Seed Tuber*

## Seed Collections

Potato is a cool-season crop, requiring low temperatures for optimum growth and tuberization. It is cultivated as a summer crop in temperate regions but is grown during the winter in the tropics and subtropics. It needs well-drained sandy loam soils. Soils rich in nutrients such as nitrogen, phosphorus, and potassium are favorable for high crop production. Potato flowers profusely and sets more fruits (berries) during long days. Both self- and cross-pollination occur, but many cultivars do not produce a sufficient quantity of pollen. However, a few set moderate to abundant amounts of pollen. Normally, potato is hand pollinated twice for better seed setting. Fruit matures in six to seven weeks after anthesis. Ripe healthy fruits are harvested for seed purposes, and seeds are separated, cleaned, dried, and stored.

*Seed germination.* Seed germination is epigeal and takes three to four weeks to complete. Seed germination is better at a constant temperature of 20°C and under diffuse light. Seed coat removal and exposing seeds to 16 to 118°C for 12 days increases germination percentage. Fresh seeds exhibit dormancy, which can be removed by after-ripening of seeds for 6 to 24 months at room temperature.

*Seed dormancy.* Fresh seeds from fully developed berries show prolonged dormancy, and seeds of early harvested berries are less dormant and show poor viability. Dormancy is beneficial, especially in delaying the germination process, and enables longer storage of seeds under ambient conditions. Seeds of tetraploid plants are more dormant than seeds of diploid plants. Dormancy can be overcome by storing seeds for seven months. The endogenous gibberellic acid level increases during the first 120 days of storage, and simultaneously the quantum of abscisic acid.(ABA) decreases. The proportion of GA and ABA regulates seed dormancy.

However, the residual dormancy persists for two to eight years in potato seeds. The external application of gibberellic acid (1,000 ppm) removes seed dormancy.

## Storage Conditions

Potato seeds are viable for relatively long periods, and they do not require specialized and expensive storage structures, as in the case of tubers. Seeds require dry, cool conditions for maintaining high seed quality in terms of viability and vigor. Seeds with 5 to 7 percent mc stored at 6 to 10°C over calcium chloride remain viable for eight years. Further, about 70 percent of seeds are viable even after ten years of storage at fluctuating temperatures (5 to 45°C). High seed viability is maintained for 10 to 13 years when seeds are stored at 0°C. At lower temperatures (0 to 5°C), seeds remain dormant for 5 to 13 years. Seeds that are low in moisture when stored in sealed containers at 1.1°C maintain high viability and vigor for nine years.

# Chapter 37
# SWEET POTATO

## INTRODUCTION

Sweet potato is also called Spanish potato. It is cultivated throughout the tropics and subtropics for its edible tuberous roots. Tubers are rich in carbohydrates and beta-carotene, and they contain a high percentage of sugars, predominantly sucrose, about 3 to 6 percent. Tubers are boiled, baked, fried, and canned. Potato flour is used in the preparation of starch, glucose, pectin, and industrial alcohol. The tender foliage, which is rich in vitamins A and C, is used as a leafy green vegetable.

### Origin and Distribution

Sweet potato originated from *Ipomoea trifida* in tropical South America. It is largely cultivated in Argentina, Brazil, Columbia, Ecuador, India, Indonesia, Japan, Malaysia, South Korea, and the Philippines.

### Morphology

Sweet potato belongs to family Convolvulaceae and has a chromosome number of 2n = 90. It is a hexaploid, dicotyledonous perennial plant with long trailing vines that is cultivated annually. Roots are tuberous, fusiform, or globular, and there are four to ten per plant. Tubers vary in size and color, ranging from white, tan, and yellow-orange to salmon orange or red. Leaves are alternate and vary in shape. Sweet potato flowers readily in the tropics.

Flowers are single and axillary; the calyx is five lobed; the corolla is funnel shaped and purple in color; stamens number five; the ovary is bilocular. Fruit is a dehiscent capsule that contains one to four seeds. Seeds are black, angular, and 3 mm long with hard testae. A large number of seeds set when tops are grafted on roots of cultivated morning glory plants.

## SEED STORAGE

Stem cuttings are normally employed for propagation. Seeds are largely used in breeding programs and for conservation of genetic diversity in the gene bank. Seeds are stored for fairly long periods under ambient conditions. They show orthodox storage behavior, can tolerate desiccation, and store well at chilling temperatures.

### Seed Collections

Sweet potato is a warm-season crop. It grows well in hot climates and needs plenty of sunshine. Low temperatures (less than 15°C) affect plant growth. Short days are congenial for flowering and root development. Sweet potato requires deep, friable, sandy loam soil. The soil should be well drained. It is cultivated through root and vine cuttings (30 to 45 cm long).

Later, it is transplanted on ridges in the main field. Crops should be provided with adequate nutrition, irrigation, and plant protection measures for healthy and optimum growth. Sweet potato is self-incompatible, and crosspollination is brought about by insects. Matured dry fruits are collected for seed purposes and are extracted for storage.

### Seed Germination

Sweet potato seeds possess a hard seed coat that delays the imbibitions process. Therefore, seed scarification with concentrated sulfuric acid is helpful in encouraging early and increased seedling emergence.

### Storage Conditions

In sweet potato, seed deterioration is greater under high seed moisture and high storage temperatures. However, low-moisture seeds stored in moisture-proof containers under cool conditions retain viability for longer periods. Jones and Dukes stored seeds at 18°C and 45 to 50 percent RH for 21 years, and 90 percent of sweet potato seeds germinated.

# Chapter 38
# EGGPLANT

## INTRODUCTION

Eggplant is one of the most popular vegetable crops in all the tropical and subtropical countries. Eggplant is also called brinjal or aubergine. The name eggplant is derived from the egg-shaped fruit. It is a popular vegetable cultivated for its edible fruits. The immature fruit is cooked as a vegetable, fried, stuffed, or boiled.

### Origin and Distribution

Eggplant is native to India, where wide genetic diversity exists. Its wild type and landraces, which are distributed in large areas, are spiny and bitter. It is largely cultivated in China, Egypt, Ghana, India, Indonesia, Italy, Iraq, Japan, Kenya, Malaysia, Nigeria, the Philippines, Puerto Rico, Spain, Syria, Thailand, and Turkey.

### Morphology

Eggplant belongs to family Solanaceae, and its chromosome number is 2n = 24. It is a weak, erect, branching, spiny, pubescent, dicotyledonous, perennial herb. It is normally cultivated as an annual. Plants are moderately tall, growing up to 1 m in height. Foliage is dark green with large thick ovate or ovate-oblong leaves. Flowers are violet colored, single or in clusters of two to five. The calyx is long, woolly, often spiny, and it is persistent on the fruit even after maturity. Petals number five and are gamopetalous; stamens number five, are yellow, and stand erect. The ovary is superior and bilocu lar. Fruits vary in size, shape, and color and have many seeds. They are oblong or obovoid; smooth and shiny; and yellow, green, white, purple, or black. Seeds are small and numerous, kidney shaped, and light brown in color. One thousand seeds weigh about 4 g.

## SEED STORAGE

Seeds are used for commercial cultivation of the crop. They are also used in genetic conservation, owing to ease of handling, are capable of maintaining genetic stability, and are inexpensive. Seeds show orthodox storage behavior. They require dry, cool conditions for better storage.

### Seed Collections

High viability and vigor are ideal qualities for seed storage. These help to withstand unfavorable storage conditions. Vigorous seeds remain viable for longer periods under both favorable and unfavorable storage conditions. Seeds should be bold, clean, and free from debris as well as from insects and other microorganisms for better storage. Quality seeds are thus produced when optimum cultural practices are followed during seed production, and these seeds can then be collected for storage.

Eggplant is a warm-season crop and therefore needs high temperature for optimum growth. Variations in temperature affect plant growth. The optimum soil temperature for seed germination is 30°C. Comparatively, seeds of long-fruited cultivars better withstand extremes of heat. The crop grows best in rich, deep, well-drained, sandy loam soils. Seeds are sown in raised nursery beds.

Healthy seedlings are transplanted in the field after four to six weeks. In high rainfall areas, crops are grown on raised beds. Optimum spacing, fertilization, irrigation, and other cultural practices should be followed.

Eggplant is a self-pollinated crop. However, cross-pollination occurs to a certain extent. Thus, an isolation distance of 50 m is provided to safeguard against contamination. Plants are checked for being true to type, especially during the flowering and fruiting stages, and any off types are removed.

Fruits are harvested about 90 days after planting. Fruit color changes on maturity and its glossy appearance is lost. Simultaneously, seeds mature and become ready for harvest. Seeds are extracted from dry ripe fruits either manually or by wet extraction, whereby fruits are crushed and the pulp and seeds are soaked in water overnight. Later, the seeds are separated, dried, cleaned, and suitably packed for storage.

### Seed Germination

Fresh seeds of eggplant do not germinate satisfactorily, requiring more time for germination, and thus show a certain amount of dormancy. Dormancy can be overcome by partially soaking seeds in potassium nitrate (4 percent) and subjecting them to an alternate temperature (20/30°C) regime. Seed treatment with gibberellic acid (100 ppm) eliminates dormancy.

Seed germination is epigeal and takes two weeks to complete. Germination is better at a constant temperature of 25°C or at alternate temperatures of 20/30°C. Kretschmer reported that seeds stored at room temperature or under low temperature for three years germinated well at 15 to 30°C, while Suzuki and Takahashi suggested that eggplant seeds require alternate temperatures for full germination. Temperatures of 30/23°C for 8/16 h, respectively, are sufficient to promote germination in fresh and oneyear- old dormant seeds. Seed germination is higher in heavier seeds than in lighter ones, and this is attributed to better seed filling. Seed weight is positively correlated to protein content and, in turn, to seedling vigor. Further, higher seedling vigor from heavier or larger seeds is attributed to better availability and metabolism of reserve food. Eggplant seeds germinate better under dark conditions. However, seeds are sensitive to salt, and germination is affected in saline soils.

## Factors Affecting Seed Longevity

Eggplant seeds can be stored fairly well under ambient conditions. High seed moisture at higher temperature affects seed longevity.

*Genetic factors.* Seed longevity varies within the genus among the different species. Certain cultivars withstand unfavorable storage conditions and remain viable for limited periods. Such cultivars also do not require special conditions for short-term storage.

*Fruit maturity.* Eggplant fruits take 30 to 40 days to mature following anthesis. Seeds extracted from fully ripe fruits give high seed quality, as seed maturity coincides with that of fruit. Mature seeds retain viability better than immature seeds.

*Fruit storage.* Eggplant fruits are stored for a week under ambient conditions. Further storage causes shriveling of fruits, affecting seed quality. Seeds from underripe fruits are particularly susceptible to freezing temperatures and become dormant; warming improves germination. Storing fruits at low temperatures or below 0°C reduces germination. Storage of mature fruits for seven days increases the germination percentage.

*Seed position in fruit.* Seeds from the basal portion of fruits give early and high germination. These seeds preserve their high quality on storage. Similarly, fruits borne on the lower portion of the plant give quality seeds and higher germination. Size of fruit and seed weight are reduced in later-formed fruits.

*Storage pests.* Quite a few insects and fungi are associated with storage of eggplant seeds. Many of the pests can be eliminated under drier storage conditions. Drugstore beetle (*Stegobium paniceum* L.) causes considerable seed damage and affects seed viability. It is controlled by fumigation with methyl bromide. The viral activity inside the seed is entirely lost after seven months of storage at room temperature, without an appreciable reduction in germination. Fungi are controlled by a combination of activated clay and carbendazim. Application of thiram or Delsan-30 protects the seeds over a period of 18 months, without affecting the germination. Seed dressing with organomercurial compounds severely reduces seed viability.

*Storage conditions.* Seed longevity is primarily dependent on temperature, relative humidity, and, to a lesser extent, oxygen. Extremes of temperature and relative humidity result in a rapid decline in seed viability and seedling vigor, and an increase in leaching of electrolytes (see Figure.1). Seed deterioration and invasions by fungi are greater at higher levels of seed moisture. The moisture is regulated by atmospheric humidity under open storage. Barton reported that seed viability was affected by high humidity during storage (76 percent RH for 64 weeks).

Seeds stored at relative humidity of 10 to 40 percent maintained 50 to 60 percent viability after five years of storage under ambient conditions. Seed moisture content of 5 to 7 percent is ideal for long storage.

## Storage Methods

Well-dried eggplant seeds maintain high viability under cooler conditions. Seeds are protected from extraneous factors, including humidity, by providing suitable moisture-proof packaging. Polyethylene bags are ideal for short storage, whereas laminated aluminum foil pouches are effective for longer storage.

*Seed storage with silica gel.* Seed storage with silica gel is beneficial in lowering the moisture content, thereby prolonging storage life. Silica gel is an inactive, inexpensive, and useful self-indicator material for seed storage. Seeds stored with silica gel in aluminum foil under ambient conditions recorded 50 percent germination after 30 months of storage, as compared to seeds stored without silica gel, which had 50 percent germination at 12 months of storage.

*Cold storage of seeds.* Seed viability and vigor preserve well under low temperatures. Mohamed, Lester, and Mumford reported that seed viability declines rapidly under warm, humid conditions. While seeds with 7 percent mc maintain viability of 92 to 98 percent at –18 to 20°C for 112 days, seeds stored in laminated aluminum foil pouches at 5°C and –20°C maintain high viability and vigor for ten years. Viability decreases to 50 percent during the second year of ambient storage.

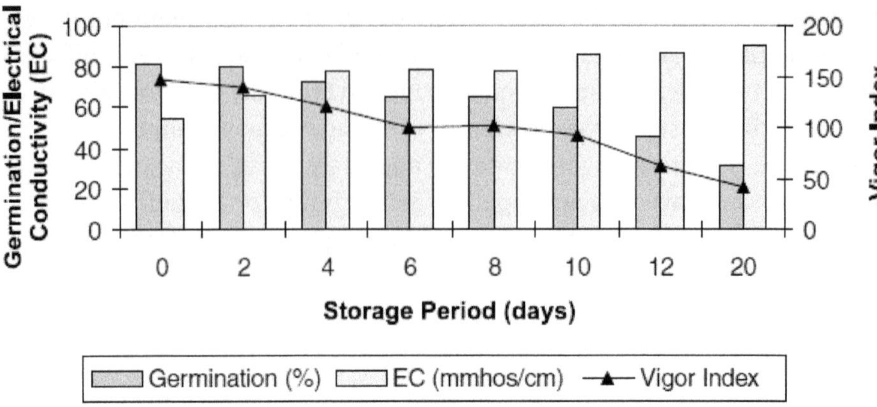


*Figure.1. Influence of High Temperatures (42°C) and Humidity (90 Percent RH) on Seed Viability, Vigor, and Leakage in Eggplant*

*Seed storage in modified atmosphere.* High atmospheric oxygen is injurious to seed viability, especially at higher levels of seed moisture. Eggplant seeds stored in laminated aluminum foil pouches along with carbon dioxide exhibited high seed germination after four years of storage under ambient conditions.

## Invigoration of Stored Seeds

Seed deterioration is a gradual and irreversible process resulting in loss of seed quality. Imposing certain treatments slows down this process. In one-year-old seeds, germination is improved by naphthaleneacetic acid (NAA) (100 ppm) application. Similarly, in five-year-old eggplant seeds, germination and vigor are improved by seed priming with gibberellic acid (50 ppm) and potassium nitrate (0.01 M).. Midstorage application of disodium phosphate to eggplant seeds is shown to improve seed germination, crop growth, and yield.

# Chapter 39

# TOMATO

## INTRODUCTION

Recent years have witnessed a dramatic rise in tomato cultivation in many areas of the world. Tomato is an important solanaceous fruit vegetable that ranks fifth in production after potato. The fruits are edible and are eaten raw or cooked as a vegetable. Ripe fruits are used in the preparation of puree, soup, juice, ketchup, paste, and powder. Green tomatoes are cooked as a vegetable and made into pickles. Tomatoes are a rich source of minerals and vitamins A and C. Seeds contain a fairly high amount (24 percent) of edible oil.

### Origin and Distribution

Tomato is native to South America. Large genetic diversity is observed within the coastal areas of Chile and Ecuador. Tomato is cultivated in the tropics, subtropics, and temperate regions. Fruits grown in cooler climatic conditions possess greater flavor and quality than those from tropical climates. Tomato is largely cultivated in Algeria, Brazil, China, Egypt, Greece, India, Indonesia, Iran, Italy, Japan, Malaysia, Mexico, Morocco, the Philippines, Spain, Turkey, and the United States.

### Morphology

Tomato belongs to the Solanaceae family and has chromosome number 2n = 24. Plant is weak and erect or climbing; it shows indeterminate, semideterminate, or determinate type of growth. It grows up to 2 m in height, and branches heavily with compound leaves. Leaves are dark green and hairy, with a characteristic odor. Flowers are yellow, perfect, and borne in clusters of four or more. Flower is composed of a short calyx tube; corolla rotate with six petals; six stamens that form on the corolla tube; and a pistil with several locules (five to nine). Fruit is a soft fleshy berry with various shapes, sizes, and colors. It may be round, pear, oblate, somewhat square, or flattened, with a smooth or ribbed surface. The skin color changes to red or yellow on ripening, but it remains green in fruits of *Lycopersicon peruvianum, L. chilense,* and *L. hirsutum.* Seeds are many, small, light brown, hairy, and kidney shaped, with a curved embryo embedded in the endosperm. One thousand seed weigh about 2.5 to 3.3 g.

## SEED STORAGE

Tomato seeds are tiny structures capable of withstanding adverse storage conditions, to a certain extent. They show orthodox storage behavior, wherein desiccation of seeds and storage at low temperatures increase seed longevity. Seeds are primarily used in propagating tomato and in breeding and genetic conservation. Farmers prefer to preserve viable seeds for short periods, such as until the next growing season, whereas breeders demand the fairly long storage for germplasm conservation. Therefore, suitable storage methods have been developed to suit these requirements.

### Genetic Factors

Tomato seeds maintain their viability for long periods, and this is attributed to genetic factors. Tomato seed longevity is primarily controlled by genetic means, and the magnitude of viability and germination potential is higher in hybrids than in inbred plants. Seed longevity differs within the genus among species and cultivars. Seed germination percentage varies among different cultivars stored for four years under ambient conditions.

### Seed Collections

Tomato is a warm-season crop and cannot survive at very hot or very cold temperatures. It prefers deep, fertile, well-drained soil. The optimum soil temperature for germination is 30°C, with a minimum temperature of 10°C. Seeds are sown in nurseries or on raised beds or ridges, and after four to six weeks, the seedlings are transplanted to the field. Seedlings are supported with stakes, especially the indeterminate types. Proper fertilization, timely irrigation, and plant protection measures should be followed. Tomato is a self-pollinated crop. An isolation distance of 50mis maintained to avoid accidental cross-pollination by insects. Plants are examined for cultivar characteristics during the flowering and fruiting stages, and the off types are removed. It takes about 60 to 90 days from planting to harvesting, depending on cultivars. Extreme temperatures, especially higher than 38°C, affect the fruit set. Ripe fruits are harvested (see Figure 1). Seeds are removed by cutting the fruits; they are then washed in water and dried. Pulp is also separated by fermentation or by using dilute hydrochloric acid. Normally, medium to large fruits are picked for seed purposes. Seeds of these fruits give higher germination and vigorous seedlings.

*Fruit maturity.* Tomato seeds reach acceptable quality on maturity of fruits. The seeds mature between 35 and 41 days after anthesis, at which time seed moisture will be about 53 to 72 percent. Seeds taken 21 days after flowering are capable of germinating but do not store long. Seed germination, vigor, and storage protein increase, and seed moisture decreases, with advances from mature green to the ripening. Fruit color changes on maturity, when fruit becomes dark red and seed dry weight and vigor reach the maximum. Yaovalak noted that seed germination and vigor are unaffected in fruits showing pale pink to dark red coloring. Seeds of mature green fruit give low germination, and rate and intensity of germination increase from the color breaker stage. Seeds attain maximum viability in the ripe stage of fruit development and show greater seedling vigor. However, fruits can be plucked at the color breaker stage for an early seed crop without hampering seed viability. Likewise, Singh, Singh, and Dhillon obtained 95 percent germination in seeds of fruits turning red and in red ripe fruits, as compared to only 64 percent in seeds of mature green fruits. Further, seed quality is affected in seeds of overripe and diseased fruits.

*Figure 1. Tomatoes Ready for Harvest*

*Fruit storage.* Storing fruit for a short period provides ample time for preparation and processing of seed material for storage. Fruits of different stages retain good seed quality for various periods. Seed viability decreases with an increase in fruit storage time. Red ripe tomatoes stored in a refrigerator (11.5°C and 57 percent RH) retain seedling vigor for six weeks and show the highest vigor during the third week of storage, as compared to the low vigor shown when stored for a short period at room temperatures (27.7°C and 75 percent RH).

*Seed drying.* Tomato seeds have 60 to 70 percent mc at harvest, and this must be reduced to 9 percent during storage for better seed quality. Seeds dried in sun and shade (alternately for 2 h) and mechanical drying with air at 40°C are optimal for safe drying of seeds for better storage.

## Seed Germination

Seed germination is epigeal and takes about one week to complete at higher temperatures. The optimum temperature is from 25 to 30°C. Tomato seeds show considerable dormancy. Secondary dormancy is induced by chilling, warm stratification, and exposure to far red light. Seed priming is reported to improve the germination percentage.

*Seed size.* Medium-sized seeds are reported to give higher germination, and seed size of about 0.8 mm is found to give higher field emergence and higher yield.

*Plant extracts.* Plant extracts contain certain growth substances that either promote or inhibit the germination process. Tomato juice inhibits germination, an effect attributed to the presence of abscisic acid. Similarly, tomato seed germination is inhibited by mandarin juice and *Emblica officinalis* fruits. Plant extracts of *Ceratophyllum demersum, Eichhornia crassipes, Marsilea minuta, Salvinia natans,* and *Spirodela polyrrhiza* stimulate germination, whereas extracts of *Cannabis sativa, Crinum asiaticum, Holarrhena antidysenterica, Nelumbo nucifera,* and *Sansevieria roxburghiana* inhibited germination.

*Fungal extracts.* Tomato seed germination is optimum in normal soils, rather than in sterilized and saline soils. Seedling growth improves with the addition of rhizosphere microflora suspension of respective plant species to the soil prior to sowing. Seed coating with spores of *Aspergillus niveus, A. rugulosus, A. tamarii, Penicillium* spp., and *Trichoderma lignorum* controls pre- and postemergence losses.

*Growth substances.* Seed treatment with growth substances such as GA, IAA, and NAA and short exposure to light and magnetic stimulus enhance the seed germination process. Seed soaking in GA4+7 in addition to osmotic priming (–1.0 MPa polyethylene glycol [PEG] 6000) increases germination of tomato seeds.

*Storage of germinated seeds.* Pregerminated and germinated seeds remain viable for shorter periods under ambient conditions. They retain their viability for five days at 5°C. Further storage reduces viability and vigor due to the degradation of enzymes and functional structures and an increase in free fatty acids. Likewise, pregerminated seeds could effectively be preserved in moist cheesecloth, with emergence percentage and emergence rate index unaffected. These seeds can also be preserved well in vacuum storage or in nitrogen at 7°C for 63 days and register excellent plant growth. Beers and Pill preserved germinated tomato seeds for five days at 0° or 5°C by suspending the seeds in two fluid drilling gels, namely Natrosol-250 and Laponite-445.

## Storage Conditions

Tomato seeds preserve well under ambient conditions. Higher temperatures and relative humidity are not favorable for storage. Seeds stored at 50°C and 77 percent RH showed slow and less germination. The optimum temperature and RH for storage are close to 0°C and not exceeding 70 percent, respectively. Low seed moisture, 5 to 7 percent, is ideal for longer storage, even if higher temperatures are maintained. Tomato seeds survive in outer space conditions for several years without adverse effects on germination, emergence, and fruit yield. Seed vigor depends on initial status of seed quality and the storage conditions. During storage, seedling vigor declines earlier and more rapidly than viability. Further, disintegration of cell membranes causes excessive leaching of electrolytes, soluble sugars, and free amino acids. The process of deterioration slows under favorable storage conditions.

*Storage fungi.* Good healthy seeds maintain their viability for longer periods. Seed health is affected under unfavorable storage conditions. Storage fungi are active at higher levels of seed moisture, causing discoloration of seeds. This is checked either by proper seed drying or by using suitable fungicidal treatment. Seed treatment with captan (2.5 g·kg −1) is effective in controlling storage fungi, and it helps retain high viability for 18 months under ambient conditions.

## Storage Methods

*Seed storage with desiccants.* Seed moisture plays a major role in deterioration. Even seed exposure to high levels of relative humidity reduces viability and vigor and affects membrane integrity.

Desiccant-like silica gel is useful in maintaining an optimum level of seed moisture. Tomato seeds stored over silica gel preserved high viability and vigor for three years under ambient conditions.

*Seed storage at low temperatures.* Tomato seeds retained high seed viability and vigor for three years under ambient conditions. Thereafter, seed viability decreased; however, the seeds were economically viable until the eighth year of storage. Barton and Garman reported that tomato seeds maintain viability for 13 years at room temperatures and for 18 years at −5°C, and stored seeds behaved similarly to fresh seeds. Seeds maintain high viability and vigor at low (5°C) and subzero (−20°C) temperatures for ten year, as compared to four years at ambient temperatures (16 to 35°C). Well-dried seeds remain viable for longer periods at lower temperatures. However, seeds should be suitably packed in moisture-proof containers to safeguard against high relative humidity and extraneous pests.

## Invigoration of Stored Seeds

Seed storage under unfavorable conditions reduces the rate of germination, uniformity, and total germination and increases the number of abnormal seedlings. Such low-vigor seeds can be reactivated by physical and chemical stimuli. Prestorage humidification and seed hydropriming impart resistance to deterioration. Seed priming with PEG 8000 at 20°C for seven days increases the rate of germination and maintains high viability at lower temperatures (10 and 20°C). Owen and Pill also reported that primed tomato seeds stored at 4°C showed maximum seed viability. Prolonged high temperature (35°C) storage of primed seeds reduces the germination rate and total germination. Repriming of primed seeds has some benefit but does not entirely reverse the detrimental effects of high-temperature storage. Further repriming is ineffective when seeds show greater loss of viability. However, a slight gain in early germination is reported.

Midstorage soaking of seeds in water or chemicals hastens the germination process by leaching certain toxic substances and activates the enzyme network involved in germination. Soaking tomato seeds in water followed by drying improves the germination and seedling vigor. Hydration is also achieved by exposing seeds to 100 percent RH for 24 h followed by redrying, promoting germination. Petrikova soaked tomato seeds in water for 24 h followed by drying at 12°C for ten days; this resulted in higher percentage of germination, higher fruit yield, and a greater number of heavier fruits.

The beneficial effects of prestorage humidification and hydropriming related to metabolic activities induced by partial hydration; the adverse effects of osmopriming were caused by a decrease in DNA repair activity due to progression in the cell cycle. Priming enhances seed germination but simultaneously lowers resistance to deterioration. Primed seeds are vigorous but have a reduced storage life. Vos, Kraak, and Bino reported that aging of tomato seeds involves glutathione oxidation. Priming reduces the oxidized form of glutathione. Hydration in conjunction with dilute solutions of sodium chloride (10–3 M) or sodium orthophosphate (10–4 M) improves field performance and productivity. Seed quality of eightmonth- old tomato seeds was improved by disodium phosphate , and that of one-year-old seeds by naphthaleneacetic acid (50 ppm.

# Chapter 40

# PEPPERS

## INTRODUCTION

Peppers are known by several names in different regions, such as bell pepper capsicum, cayenne, chilli, paprika, red pepper, and sweet pepper, according to their shape, color, and pungency. Immature and mature fruits are eaten raw as salad vegetables, and various vegetable dishes are made from them. Sweet peppers are nonpungent, have a mild flavor, and are thus widely used in salad. Ripe chilli fruits are pungent, and their dry powder is commonly used as spice in cooking various foods. Hungarian type paprikas are more pungent, whereas Spanish and European types are less pungent.

The pungency in the fruit is due to capsaicin (C18 H27 NO3) content, which is mainly confined to the septa and placental tissue. Peppers are a rich source of vitamins A and C.

### Origin and Distribution

Peppers originated in Central and South America, more particularly in Mexico and surrounding countries. It is widely grown in the tropics, subtropics, and temperate regions. Some of the leading pepper-growing countries are Algeria, Argentina, Bulgaria, China, Hungary, India, Indonesia, Japan, Kenya, Mexico, Malaysia, Nigeria, the Netherlands, the Philippines, Spain, Sudan, Thailand, Turkey, the United States, and Yugoslavia.

### Morphology

Peppers belong to family Solanaceae, and their chromosome number is 2n = 24. They are short-lived, dicotyledonous, perennial plants in the tropics, but they are cultivated as an annual. Plants consist of a strong root and shoot system and grow up to 1.5 m in height. Plant stems are hard and branched, with simple, ovate, light green to dark green leaves. Flowers are small, axillary, and single or two to three in cluster at the node; the calyx is five lobed; the corolla is white or greenish, campanulate, and five petaled; the five stamens have bluish anthers; the ovary is two celled. Fruit is a podlike berry, erect or drooping, indehiscent, and many seeded. It varies in shape (heart, cylindrical, round, cherry, long and thin, or square), size, and color. Initially, fruits are green or purple in color and later become red, orange, yellow, brown, cream, or purplish on ripening. Seeds are somewhat circular, yellow, light in weight, and 3 to 5 mm long. One thousand seeds weigh about 3.5 to 5 g, depending on variety and growing conditions.

## SEED STORAGE

Pepper seeds show orthodox storage behavior. Seeds are predominantly used in crop production and for genetic conservation. Seeds are genetically stable in long-term conservation. Seeds remain viable for two to three years in ambient conditions. High atmospheric humidity and high temperatures shorten the storage life of seeds. The germinating energy in stored seeds decreases more rapidly than germination capacity, and it is lost completely by three years when stored at room temperatures.

The seed storage life is primarily dependent on genetic factors. Seed longevity differs among the different cultivars during storage and is influenced by storage temperatures and cultivation condition. Some genotypes withstand aging to a certain extent and remain viable for relatively longer periods under ambient conditions. Such genotypes produce improved seed longevity in hybrids when used as parents.

### Seed Collections

Peppers grow well in warm, humid climates. High temperature induces flowering, while excessive rainfall damages the crop, affecting the fruit set and causing fruit rotting. Peppers prefer deep, fertile, well-drained soils. Seeds are sown in nurseries on raised beds and subsequently transplanted to the field four to six weeks after sowing. In the field, adequate spacing, fertilization, weeding, and suitable plant protection measures must be followed to get a healthy crop. Though peppers are a self-pollinated crop, about 5 to 40 percent are cross-pollinated. Normally, bees, thrips, and ants cause cross-pollination. To prevent pollen contamination, an isolation distance of 400 m is adopted. Plants are inspected for vegetative, flowering, and fruiting characters during the flowering and fruiting stages, and off types are removed.

Fruit matures four weeks after anthesis. Red ripe and disease-free fruits are plucked for seed purposes. Seeds are extracted manually or by machine, the latter giving better recovery at less expense.

*Fruit maturity.* Mature bold seed maintains high seed quality during storage. Seed maturity coincides with that of fruit maturity and is expressed by change in fruit color. Early harvested fruits give poor-quality seeds with low viability and vigor. Fruits harvested at color breaker stage have high viability. Metha and Ramakrishna reported that seeds extracted from fruits 48 days after anthesis have higher viability and vigor. Further, seeds extracted from second and third fruit pickings retain high viability after 12 months of ambient storage.

Seeds are fully capable of germination and impart tolerance to desiccation just before or at mass maturity. The maximum potential longevity occurs after 10 to 12 days of mass maturity. Seeds harvested at the ripe stage show higher germination and high seedling vigor , and can be stored longer under ambient conditions . Sanchez and colleagues noted that seeds extracted from mature green fruit did not germinate and required 14 days of storage for germination. Seeds from red fruits (50 days after anthesis) and overmature red fruits (60 days after anthesis) possessed higher germination capacity and greater dry weight. Seeds are generally allowed to mature for a short period within the fruits after harvest for better seed quality.

*Fruit storage.* Dry pepper fruits can be stored for various periods, depending on the climatic and storage conditions. Fruit seeds stored up to six months exhibited higher germination. Pepper fruits preserve well in cloth bags or paper bags, as these allow free exchange of gases and water vapor from the fruits. Seeds of these stored fruits gave 75 percent germination for three years, whereas seeds of fruits stored in sealed conditions in aluminum foil did not germination.

*Seed position.* Seed viability is influenced by the position of seeds inside the fruit. Seeds from the basal portion of the fruit have higher potential for viability and vigor as compared to those from the middle and tip.

## Seed Germination

Fresh seeds do not germinate readily and show the presence of dormancy. Seeds require six weeks of after-ripening to germinate. Several by-products, such as alcohol, aldehydes, ketones, esters, hydrocarbon, and furans, are released from seeds during after-ripening. Pepper seeds germinate well at a constant temperature of 25 to 35°C and are inhibited below 20°C, which can be overcome by seed priming (PEG 6000 or potassium nitrate [KNO3] + potassium orthophosphate [K3PO4]). Seed germination is epigeal and takes about six to ten days to complete. Seeds of cultivars derived from other species, such as *Capsicum baccatum, C. chinense,* and *C. pubescens* do not germinate at a constant temperature and require an alternate temperature regime. Pepper seeds are sensitive to salinity, and germination is affected by high levels of salts, more than 4,000 ppm. Seed germination is rapid and higher when a little fruit pulp is retained along with the seed during drying. This aids better translocation of metabolites from pulp to seed at higher temperatures. Seed germination is enhanced by the use of plant stimulants such as atonik, and such treated seeds can also be stored for longer periods. Application of growth substances such as GA and IAA improves seed germination.

*Storage of germinated seeds.* Germinated seeds are more perishable and do not store long. However, when such seeds are placed in polyethylene bags (0.025 mm) with a vacuum seal or in nitrogen at 7°C they remain viable for 63 days.

## Storage Conditions

A high seed quality is essential for longer storage of seeds, and seed losses are minimal under ideal storage conditions. Initial seed quality, especially high vigor, contributes to longer storage life. Prevalence of high relative humidity and temperature shorten storage life. Chemical composition of seeds and atmospheric relative humidity determine the seed moisture content, thereby deciding seed longevity. The low level of seed moisture is maintained and protected from atmospheric humidity by suitable packag- ing. Sundstrom reported that 10 percent moisture is optimum for better storage of *Capsicum frutescens* seeds. In a study conducted by Popovska, Madenovski, and Mihajlovski , pepper seeds (8 percent mc) were stored at 10.7 to 23.8°C and 45.7 to 85.7 percent RH for eight years in cloth bags, glass containers, polyethylene boxes, plastic bags, and tins. Among all, seed viability was highest in glassware-stored seeds, while in others, viability was lost gradually, and more rapidly in seeds stored in cloth bags. In a study by Fischer , seeds remained viable for seven years when packed in sealed jars or plastic bags and stored at 5 to 10°C.

However, seed lost viability rapidly under unsealed conditions. Under fluctuating temperatures from 2 to 26°C, seed viability decreased from 92 to 29 percent during five years of storage. Ultra-low-temperature storage of seeds at –70°C did not improve the storability over that of –12°C. On aging, seeds became brown and lost vigor and viability. Such seeds lose more electrolytes and soluble sugars on imbibitions through leaching.

*Seed storage at low temperatures.* Pepper seeds preserve well at low temperatures. Seed deterioration is less at 5°C storage. High seed viability and vigor were maintained in seeds stored at 5 or –20°C after five years of storage, as compared to three years only when seeds were stored at ambient temperatures. In bell pepper, too, high seed viability was maintained at low (5°C) and subzero (–20°C) temperatures for fifteen years, while under ambient conditions, bell pepper seeds remained viable for two years . The loss of viability was associated with greater leaching of electrolytes, soluble sugars, and free amino acids from the seeds (Table.1). Seeds with 4 percent moisture stored at 32.2°C maintained viability for at least three year.

Further, seeds preserved viability for five years at 5°C and for ten years at –4°C in sealed containers. Furthermore, high seed viability was maintained in laminated aluminum foil pouches stored at 5 and –20°C for ten years..

*Modified atmospheric storage of seeds.* Seeds stored in partial vacuum under ambient conditions remain viable and vigorous for three years. A higher concentration of carbon dioxide inhibits germination and preserves viability for five years under ambient conditions.

*Storage fungi.* Pepper seeds associate with certain fungi under conditions of high seed moisture and high atmospheric RH. *Alternaria, Aspergillus, Chaetomium, Colletotrichum, Fusarium, Penicillium,* and *Rhizopus* species affect seed quality during storage, and these are effectively controlled with fungicidal applications. Fungicides, such as captan, carbendazim, copper oxychloride, and mancozeb are effective in maintaining good seed health. Sodium hypochloride and captan at 2.5 percent give effective control of storage fungi and high seed germination during storage. The *Ocimum* oil is more effective than synthetic fungicides, and it does not show any adverse effect on seed germination or seedling growth during storage.

## Invigoration of Stored Seeds

Seed loses germinating capacity during storage. Such low-vigor seeds regain their quality to a certain extent when treated with growth substances or other physical stimuli. Seeds of cv. California Wonder showed improved germination after soaking in water for 72 h followed by gibberellic acid (50 ppm) application. Likewise, seed quality improved with the application of potassium dihydrogen phosphate (3 percent) to four-year-old seeds , whereas, repriming of primed seeds stored at 5°C caused reduction in seed germination.

*Table.1. Seed Viability and Leakages During Storage of Bell Pepper Seeds repriming of primed seeds stored at 5°C caused reduction in seed germination.*

| Storage Temp. (°C) | Germination (%) | Dry Weight (mg) | Electrical Conductivity (mmhos) | Soluble Sugars ($mg{\cdot}g^{-1}$) | Free Amino Acids ($mg{\cdot}g^{-1}$) |
|---|---|---|---|---|---|
| Room | 0 | — | 1,493 | 19.2 | 8.4 |
| 5 | 96 | 2.1 | 562 | 2.3 | 1.5 |
| –20 | 97 | 2.6 | 674 | 4.5 | 2.3 |

# Chapter 41

# OKRA: *ABELMOSCHUS ESCULENTUS* (L.) *MOENCH*

## INTRODUCTION

Okra is commonly called gombo, gumbo, and lady's finger. It is cultivated for its edible pods. The immature pods are boiled, fried, or cooked as a vegetable. Tender fruits are also canned or dehydrated. Plant extract is used for clarifying sugarcane juice in jaggary preparation. Sometimes fiber is extracted from the stem and used in the textile industry. Pods are rich in vitamins and minerals such as calcium and potassium. Seeds are rich in oil and contain about 20 percent edible oil.

### Origin and Distribution

Large genetic diversity in okra is found in eastern Africa. Okra is presumed to have originated in Ethiopia, and the wild forms are distributed in Ethiopia and Sudan. Okra is grown in the tropics, subtropics, and, during summer, in temperate regions. It is cultivated in China, Egypt, India, Indonesia, Malaysia, Nigeria, the Philippines, Puerto Rico, Sudan, and Trinidad.

### Morphology

Okra belongs to the mallow family Malvaceae, and its chromosome number is 2n = 72 and 144. Plant is an erect, woody, dicotyledonous annual with well-developed taproots. Shoots grow up to 1 to 2mheights. Stems are green or tinged red; leaves are broad, alternate, hairy, three to five lobed, and cordate. Flowers are single, perfect, axillary, yellow, bell shaped, with epicalyx, five sepals, and five petals; stamens are many, and the ovary is superior. Fruit is a capsule that dehisces longitudinally on drying. It is light to dark green or purple, round or ridged. Seeds are round, medium, and dark green, gray, or black, and 30 to 80 seeds are present in one pod. One thousand seeds weigh about 50 g.

## SEED STORAGE

Okra is commercially cultivated through the seeds. Its seeds show orthodox storage behavior, in which the longevity improves with desiccation and storage at chilling temperatures.

### Seed Collections

Okra grows best under warm conditions. The optimum temperature for germination is 25 to 35°C, and no germination occurs below 16°C. Rains during fruit development and maturity affect seed quality. It needs fertile, well-drained soil. Seeds are directly sown on the ridges in the field. Regular weeding, irrigation, fertilizer application, and plant protection measures are followed. Plants are self-pollinated, but about 20 percent cross-pollination occurs by insects. Isolation distance of 400 m is maintained to avoid pollen contamination. Plants are inspected for various morphological characteristics during the flowering and fruiting stages, and off-type plants are removed. Flowering starts 35 to 60 days after planting (see Figure.1). Fruits turn from green to gray or brown and become hard on ripening. Seeds obtained from basal pods are heavier and of good quality. Seeds separated from dried pods are cleaned, graded, and dried before storage.

*Fruit position.* Seed viability and vigor are affected by the fruit's nodal position. Seed quality is better in seeds collected from fruits borne at the third to sixth nodes. Seeds from fruits borne up to the seventh node showed higher germination and seedling vigor. Seed longevity decreases with an increase in fruit nodal position (see Table.1).

*Fruit maturity.* Fruit fiber content increases with advancing maturity. Fruit becomes hard and turns gray or brown. Seed moisture is 85 to 90 percent at earlier stages of seed development, and it decreases to 19 to 24 percent on seed maturity. Reduction of seed moisture and greater dry weight accumulation are rapid at 30 to 35 days of anthesis. Seeds up to ten days of development do not germinate; thereafter, germination increases. Seed attains maximum viability and vigor between 30 to 35 days after fruit development. When a crack appears on the fruits, they are ready for harvest. Further, delay in fruit harvesting affects seed quality. Demir observed that maximum seed dry weight occurs at 31 days after anthesis, with a moisture content of 71 percent on a dry weight basis, but high seed quality in terms of germination was recorded 52 days after anthesis, with seed moisture content at 12 percent and fully matured seeds.

*Figure.1. Okra Seed Production Plot*

*Table.1. Nodal Position of Fruit and Seed Germination (Percent) During Ambient Storage of Okra Seeds*

| Nodal Position | Initial Germi-nation | 1st Year of Storage | 2nd Year of Storage | 3rd Year of Storage | 4th Year of Storage | 5th Year of Storage |
|---|---|---|---|---|---|---|
| 3 | 97 | 98 | 95 | 93 | 83 | 29 |
| 4 | 97 | 100 | 99 | 97 | 87 | 28 |
| 5 | 98 | 97 | 96 | 97 | 83 | 30 |
| 6 | 98 | 99 | 93 | 95 | 83 | 22 |
| 7 | 94 | 98 | 95 | 97 | 82 | 18 |
| 8 | 94 | 93 | 88 | 95 | 72 | 12 |
| 9 | 94 | 92 | 83 | 91 | 72 | 11 |
| 10 | 95 | 90 | 80 | 75 | 65 | 0 |

## Seed Germination

Seed germination is epigeal and takes five to seven days to complete. However, it is delayed due to the presence of a hard seed coat. Seed germination is better at a constant temperature of 25 to 35°C or at alternate temperatures of 20/30°C for 16/8 h, respectively.

*Seed dormancy.* The seed coat is somewhat hard, and the embryo develops slowly during germination. High-moisture seeds tend to show little or no dormancy. Reduction of moisture to 4 to 6 percent causes hard-seededness, which leads to slow uptake of water. To overcome this, seeds are scarified for higher germination. Likewise, seed germination at alternate temperatures (20/30°C for 16/8 h) removes the dormancy.

Further, exposing seeds to 40°C for four days increases the water uptake and total germination. Use of a growth regulator such as gibberellic acid (50 ppm) improves the germination percentage and total yield. Seed exposure to physical stimuli such as a 40 megahertz (MHZ) radio frequency electric field reduced the number of hard seeds and gave higher germination. Dutta recorded the stimulating effects of gamma rays on okra seed germination up to 60 kr dosages. Beyond this dose, germination percentage was reduced.

*Seed color.* Okra seed color ranges from green to gray and black. Black seeds show poor seed quality. Seed germination and seedling vigor are rapid and higher in green than in black seeds.

*Seed size.* Seed size varies in okra fruit. Seeds from the peduncle end are larger, heavier, and bolder than those from the apical end. Large seeds show better germination than small seeds after 12 months of ambient storage (27°C and 67 percent RH).

## Storage Conditions

Okra seeds preserve well under ambient conditions. High storage temperature and high relative humidity reduce the storage life of seeds. Therefore, dry, cool conditions are to be created for longer storage of seeds.

*Seed moisture.* Low seed moisture, preferably 5 to 7 percent, is ideal for seed storage. Agrawal stored okra seeds (10 percent mc) for 37 months in ambient conditions. Moisture content of 13.6 percent or less is ideal for maintaining high seed viability for short periods, while 8 percent moisture maintains viability for seven years at room temperature. Further high-moisture seeds (19 and 25 percent) failed to germinate after one year of storage at room temperature, and after five years in cold storage. Reduction of seed moisture slows down the deterioration process. Seeds with 12 percent or less moisture stored in cold storage retained 90 percent germination capacity for 11 years.

*Storage temperature.* Low storage temperature is congenial for longterm storage. However, high-moisture seeds are not suitable for sealed storage at low temperature. High seed viability and vigor are maintained for ten years at 5 and –20°C, as compared to three years at room temperatures . Seed viability rapidly decreases at higher storage temperatures, which promote the leaching of electrolytes and solute sugar from the seeds.

*Seed packaging.* Packages protect seeds from various extraneous factors, such as high relative humidity, insects, and fungi. Seeds are commonly stored in polyethylene bags, cloth bags, glass containers, laminated aluminum foil pouches, or aluminum cans, depending upon storage period required, cost of seed material, and storage conditions. Cloth and paper bags are generally suitable for open storage for shorter periods, and moistureproof packages are used for longer storage at low or subzero temperatures.

Polyethylene bags are effective in maintaining high viability at 5°C, and laminated aluminum foil pouches at –20°C. Seeds stored in glass containers at room temperature and in cold storage retained germination capacity for 29 months. Even seeds stored in polyethylene bags show similar effects, unlike those stored in paper bags, which showed low germination percentage.

*Air.* High levels of oxygen and moisture promote metabolism and shorten the storage life of seeds. High-moisture okra seeds stored in a high level of oxygen and a low level of carbon dioxide showed poor germination, while high carbon dioxide and low oxygen levels tend to maintain seed viability.

*Storage fungi.* Storage fungi affect seed quality and give low germination. Normally, fungi grow at high levels of seed moisture. Okra seeds associate with fungi such as *Actinomucor* sp., *Alternaria tenuis, Aspergillus flavus, A. niger, A. sulphureus, A. terreus, Cephalosporium curtipes, Cehaleospora irregularis, Fusarium moniliforme, Gonatorrhodiella* sp., and *Rhizopus nigricans.* High seed moisture encourages the growth of *Botrytis* spp. at room temperatures of 21.1 to 37.8°C and *Penicillium* spp. both at room temperatures and in cold storage (1.7 to 4.4°C), thereby reducing germination. High seed germination can be maintained for 25 months by treating seeds with captan (2 grams per kilogram [g·kg –1]).

## Invigoration of Stored Seeds

Seed deterioration causes a decline in germination and vigor and is influenced by the chronological age of the seeds. Germination, vegetative growth, reproductive growth, and yield can be increased by the application of gibberellic acid (50 ppm) to aged seeds and by the application of indoleacetic acid (100 ppm) to one-year-old seeds.

# BULB CROPS

# Chapter 42

# ONION: *ALLIUM CEPA* L.

## INTRODUCTION

Onion is a popular bulb vegetable crop in temperate, tropical, and subtropical regions. Both immature and mature bulbs are eaten raw, cooked, or used in the preparation of different vegetable dishes. The green leaves are also used in salads, and cooked as a vegetable. Onion contains high amounts of carbohydrates, calcium, phosphorus, vitamins A and B, and the volatile substance allyl propyl disulphide. Onion has medicinal value in that it possesses antibacterial properties.

### Origin and Distribution

Onion originated in Central Asia comprising Afghanistan, Iran, Tajikstan, Turkey, and West Pakistan. It is cultivated in Argentina, Brazil, China, Colombia, Egypt, Ghana, India, Indonesia, Iran, Italy, Japan, Korea, Malaysia, Mexico, Morocco, Myanmar, the Netherlands, Nigeria, Pakistan, the Philippines, Poland, Russia, South Africa, Spain, Turkey, and the United States.

### Morphology

Onion belongs to the family Alliaceae and has chromosome number of 2n = 16. It is a monocotyledonous biennial herb but is cultivated as an annual. Stems are short with poorly developed root systems, and adventitious roots emerge from the base of the stems. Leaves are alternate, hollow, and arranged in a circle, forming a bulb. Bulbs are flat, globular, or cylindrical in shape. Bulb color varies, being either white, yellow, brown, pink, red, or green. Inflorescence is a terminal umbel consisting of 50 to 2,000 tiny flowers.

Number of seed stalks varies from 1 to 20, depending upon bulb size, cultivars, and growing conditions. Flowers are white, with six stamens and a protoandrous superior ovary, and trilocular. Fruit is a capsule that splits longitudinally and contains one to two seeds. Seeds are black, shiny, and wrinkled. Embryo is crescent shaped or curved and embedded in the endosperm. One thousand seeds weigh about 3.5 g.

## SEED STORAGE

Onion is propagated by seeds. Seeds are short-lived under ambient conditions. Seeds are also used for evolving new varieties in breeding programs, and for long-term conservation of genetic diversity. Onion seeds exhibit orthodox storage behavior; they can withstand loss of moisture and have extended longevity under low-temperature storage conditions. Seed longevity of several crop species was studied under different atmospheric conditions; and onion seeds were found to be the most short-lived. Ellis and Roberts also reported that onion seeds lost viability more rapidly than seeds of other crops. Onion seeds preserve viability for 6 to 12 months under ambient conditions.

Seed longevity is primarily governed by genetic means. However, it is also regulated by storage conditions. It varies in different genotypes, species, and cultivars. Good-storer cultivars withstand relatively higher temperatures and relative humidity. Doijode identified certain good and poor-storer cultivars. Poor-storer types showed greater cellular damage during storage, as evidenced by excessive leakage of electrolytes and soluble sugars. These cultivars are useful in studying inheritance of seed storability and in hybridizing for better storage quality.

### Location of Seed Storage

Extreme climatic conditions affect seed longevity. Areas where dry and cool climates prevail during most of the year are suitable for safe seed storage. Agrawal identified the places suited for short-term seed storage based on prevailing temperatures and relative humidity during a year. Harrington (1960b) suggested that the combination of temperature (°F) and relative humidity (percent) should not exceed 100 for better storage.

### Seed Collections

Onion is a cool-season crop that requires low temperatures for bulb formation and flowering. It withstands frost. Plants grown under low temperatures (15 to 16°C) yielded more seeds than those grown under high temperatures (22 to 23°C). Heavy rainfall areas are not suited for cultivation. Deep, fertile, sandy loam soil with good drainage is ideal for onion cultivation. Onion has a shallow root system and needs frequent irrigation.

Onion is propagated by seeds. Seeds are sown in a nursery, and fourweek- old seedlings are transplanted to the field. Soil sterilization is useful, especially in nursery bed preparation, for reducing pink root rot disease and for higher seedling vigor. Seeds are produced by the bulb-to-seed or seed-to-seed method. In the former method, mature healthy bulbs are planted for seed production. Bulbs are stored for four to five months before planting for curing. Bulbs stored at 10°C for 90 days flowered earlier, produced more flowers, and had the highest seed yield.

Likewise, bulbs stored for 12 weeks at 10°C resulted in higher yield. Onion yield increases with increased seed size. Onion is a cross-pollinated crop, and pollination is mainly brought about by bees (see Figure.1). An isolation distance of 1,600 m is used to maintain genetic purity. Plants are adequately fertilized, and protected against pests and diseases. Plants with double bulbs and thick necks are removed to maintain varietal quality. Heads are collected when fruits open and expose the black seeds (see Figure.2). Later they are cured for two to three days and then threshed; seeds are separated, cleaned, dried, and suitably packed for storage.

*Seed maturity.* Bold, uniform, healthy, and high-vigor seeds are best suited for storage. Harrington opined that seeds attained maximum dry weight during maturity. These mature seeds store longer than immature seeds. Sandhu, Nandpuri, and Thakur observed that seeds harvested one week prior to full maturity gave normal percentage of germination, while germination decreased in seeds harvested two weeks early. Further, a delay in harvesting affects seed quality by risking rain damage or seed shattering.

*Figure.1. Onion Plants Protected from Insects*

*Figure.2. Onion Heads Ready for Harvest*

*Seed drying.* Seeds are capable of exchanging moisture with the atmosphere. The process is faster in seeds rich in storage proteins, and it is slower in onion seeds. Seeds absorb moisture from the atmosphere when the vapor pressure is lower in seeds, and they lose moisture when the vapor pressure is lower in the atmosphere. When the two vapor pressures are equal, no movement of moisture vapor occurs, and the system is in equilibrium. Seeds dry on the movement of water vapor out of the seeds into the atmosphere by creating a moisture gradient. Seed structure will be damaged at a higher rate of evaporation. Thus, seeds are preferably dried at low temperature and low humidity. Drying seeds with dehumidified air at 25 to 40°C did not affect the germination percentage. Seeds are dried by various means, such as natural drying; sun drying; heated, unheated, and dehumidified air drying; drying with desiccants; and vacuum and freeze-drying. Each has advantages and disadvantages. Seeds are protected against heat damage during drying. Seeds dried in scorching heat and sun showed reduced germination. Copeland noted that drying too rapidly at a high tempera- ture was injurious to seed life. The reduction of seed moisture from 10.0 to 6.5 percent is more beneficial for preserving high viability than nondrying of seeds. Drying with silica gel is optimum for maintaining high seed viability for a longer period. Earlier, Barton believed that onion seeds should be dried to one-third to one-half of their original moisture content for satisfactory storage at room temperatures.

There are different opinions about ultralow-mc drying of seeds: It damages the cell structure and contents, thus affecting viability and seed storability, and also it is unsuitable for storage of large quantities of seeds. Freeze-drying for one day increases the seed longevity. Freeze-dried seeds showed 61 percent germination after one year of storage at 50°C in sealed containers. Further excessive drying is to be avoided, as water in the protective monolayer should not be removed.

Furthermore, freeze-dried seeds (4.1 percent mc) stored in desiccators over zeolite at 40°C gave 79 percent germination after six years of storage. Freeze-drying of seeds improves storability at higher temperatures. Seeds with 2.0 to 3.7 percent (ultradry) moisture preserve higher viability than those with 5.5 to 6.8 percent (dry) at 20°C storage, while no difference was observed at –20°C storage. It appears that ultradrying of seeds is useful, especially for seed storage under ambient conditions.

*Seed moisture.* High seed moisture is injurious to the storage life of seeds. The rates of seed deterioration and fungal damage are greater at higher moisture contents. Seed moisture is higher at the initial stages of seed development, and it decreases on maturation and ripening. Seed moisture content depends on seed composition, air temperature, and relative humidity.

At any given moisture level, seed viability and vigor decrease more rapidly at higher temperatures. Seedling vigor exhibits a rapid decline earlier than seedling viability. Harrington  reported that the seed longevity period doubled for every 1 percent decrease in moisture content. Low-moisture seeds are viable for longer periods. According to Rocha , loss of seed viability is rapid between 13 and 15 percent moisture content and slower at 11 percent. Low moisture decreases the seed deterioration, as it is unsuitable for the growth of pathogens.

## Seed Germination

Seed germination is rapid and high in onion at 20°C, whereas in another species, *Allium unifolium,* germination was high at 5°C, and seeds did not germinate at 15°C. Fresh seeds show a certain amount of dormancy. According to Lovato and Amaducci , after-ripening of two months is required to eliminate the dormancy. Thus, dormancy tends to delay germination rather than completely prevent it. The cotyledon tip elongates during germination, carrying the radicle and plumule out of the testa.

The elongated cotyledon forms a sharp bend that pushes to the soil surface, becomes green and straight, and produces foliage. Usik  reported that large seeds give better field emergence than small seeds, and their yield also is higher by 18 percent. Higher concentrations of carbon dioxide inhibit germination.

## Storage Conditions

Storage conditions such as temperature, relative humidity, oxygen, light, and pathogens contribute to seed deterioration.

*Temperature.* Seed longevity decreases with an increase in storage temperatures. According to Harrington , seed longevity decreased by half for every 5°C rise in storage temperature. The temperature is closely related to seed moisture, and the combination of these two factors seriously affects seed deterioration. High-moisture seeds at higher temperatures deteriorate rapidly. Szabo and Viranyl reported that onion seeds lose viability faster under variable temperatures and relative humidity. Kurdina noted a greater decrease in seed viability at 20°C and 80 percent RH.

High-vigor seeds tolerate such effects. Different workers have reported various longevity periods at room temperatures. Thomazelli, Silva, and Sediyama observed that onion seeds are viable for 60 days or less under ambient conditions, and Bosewell et al. also noted a shorter storage life for onion seeds, which remained viable for less than 60 days at 26.7°C and 75 to 80 percent RH. According to Amaral and Bicca , onion seeds remained viable for 12 months in an aluminum can, whereas Bacchi reported that seeds remained viable for four years at room temperatures when relative humidity was less than 10 percent. Seeds having 6.2 percent moisture remained viable for 13 years at room temperatures. Further, the reduction of seed moisture to 1.2 percent is beneficial for longer storage life (nine years) at room temperatures. In another study, seeds having 6 to 8 percent moisture, packed in polypropylene film or triple-laminated aluminum foil, and stored at 22°C and 40 to 60 percent RH remained viable for 65 months.

Seed deterioration is rather slow at lower storage temperatures, which are also noncongenial for insect and pathogen activities. Seed viability was lost more rapidly under higher storage temperature in high-moisture seeds than in low-moisture ones. In another study, seeds with 8 and 15 percent mc were packed in muslin cloth and in aluminum foil and stored at 5 and 30°C.

Seeds in muslin cloth experienced moisture loss from 15 to 7 percent under high temperature and preserved high seed viability, but seed germination was zero within one week of storage in high-moisture seeds packed in aluminum foil pouches stored at 30°C. Therefore, high-moisture seeds should be stored either in moisture-pervious containers or in unsealed conditions, especially under ambient storage conditions. Horky maintained onion seed viability for seven years by storing them at 0°C.

Subzero (–20°C) temperatures are most suitable for longer storage. Doijode preserved onion seeds for 15 years at 5 and –20°C. A higher percentage of germination was recorded in seeds stored at –20°C than in those stored at –5°C, particularly when seeds were stored in laminated aluminum foil pouches. Seedlings that emerged from stored seeds did not show any abnormalities, and they were on par with seedlings from fresh seeds for seed quality and morphological traits. Seeds stored for ten years at 5°C showed reduced germination percentage from 94 to 68 percent. Stanwood and Sowa  noted greater seed deterioration at 5°C than at –18 or –196°C. Seeds stored at higher temperatures (40 to 50°C) and relative humidity (75 percent) showed certain unacceptable changes in seedlings, such as higher frequency of chromosomal aberrations. The disturbance of genetic stability was associated with loss of seed viability.

*Relative humidity.* Humidity around the seeds regulates the seed moisture. Dry seeds gain moisture rapidly in high humidity conditions. Further, a high level of humidity favors the growth of storage fungi. When Bosewell and colleagues  stored seeds at 44, 66, and 78 percent RH, percentage of germination was 73, 37, and 0 after 250 days of storage. Likewise, Barton  reported that onion seeds lost viability rapidly at 55 and 70 percent RH, and they preserved high viability at 35 percent RH. Humidity fluctuation from 55 to 76 percent is more injurious to seed viability. To prevent the ill effects of high humidity, seeds were stored at a low humidity level and suitably packed in moisture-proof containers. Seeds stored at 10 percent RH gave 46 percent germination after five years of storage at 20°C, and they lost viability rapidly at RH of 40 percent and above . Rusev and Bacvarov  suggested storing dry seeds in sealed containers at room temperature with the enclosed air saturated with acetylene.

*Air.* The removal of air from storage containers enhances seed longevity. Seeds stored in partial vacuum gave 54 percent germination, as compared to those stored in air for 18 months under ambient conditions, which showed 27 percent germination , while no difference was noted when seeds were stored in air or partial vacuum at 34 to 40°C for 62 months. According to Lougheed and colleagues , the deleterious effects of high relative humidity decrease when onion seeds are stored under reduced atmospheric pressure for 14 weeks, and the moisture absorption is reduced by packing seeds in partial vacuum.

*Light.* Light hastens the germination process in certain vegetable crops, but darkness increases the germination percentage in most of the vegetable crops. This is attributed to the role of phytochrome. A 2 h exposure of seeds to fluorescent light every day prevents germination. The same seeds could be made germinable by subjecting them to dark conditions.

Thanos and Mitrakos  believed that intermittent light prevents the germination process more than continuous light, and inhibition is greater at higher light intensity. This is beneficial, especially in preserving seeds for a shorter period, because it inhibits sprouting during unfavorable storage conditions.

*Storage fungi.* Seeds are carriers of several pathogens on their surface and inside their tissue. The growth of these organisms is enhanced by the presence of high seed moisture and high relative humidity. Kurdina observed a greater number of mold fungi colonies at relative humidity above 70 percent and at 20°C storage. These pathogens cause seed discoloration, produce much of the heat in seeds, and reduce the germination percentage. The fungi associated with onion seeds are *Alternaria alternata, Aspergillus flavus, A. niger, A. ochraceous, Chaetomium globosum, Curvularia lunata, Drechslera australiensis, Fusarium* spp., *Penicillium cyclopium, Rhizopus stolonifer,* and *Stachybotrys atra.* The production of fungal metabolites reduces seed viability and vigor during storage. Use of fungicides effectively controls storage fungi. Gupta, Chhipa, and Bhargava reported that onion seeds remained viable (49 percent) up to a period of 40 months when treated with Ceresan and thiram (0.2 and 0.3 percent) followed by mancozeb (0.2 percent). Maintenance of low seed moisture (8 percent), low temperature (10°C), and RH less than 70 percent is beneficial for longer retention of pathogen-free seeds.

## Seed Storage Methods

Different storage methods, such as the use of desiccants, low temperature, and modified-atmosphere storage, maintain high viability and vigor for various periods. The methods should consider the principle factors associated with seed deterioration, such as temperature, humidity, oxygen, and pathogens, during storage. The choice of methods depends on the period of seed storage, kind and quantity of seed material to be stored, and cost. Maintaining genetic stability is the most important factor, apart from the retention of high viability and vigor in seeds. Seed storage begins with selection of bold, well-matured, healthy, and vigorous seeds for storage.

*Storage with desiccants.* In this method, seeds are stored in moistureproof containers, such as glass bottles, plastic boxes, or aluminum cans or pouches, along with a known quantity of desiccant, such as silica gel. The desiccant absorbs excess moisture and maintains a low moisture level throughout the storage period. On moisture absorption, blue silica gel turns white. Drying in the sun or in an oven further dehydrates it. Silica gel has added advantages; it is easy to use, less expensive, and more effective in maintaining low seed moisture. According to Kruse , silica gel should be added at the rate of 20 percent of seed weight for each 1 percent reduction in moisture content, while Grubben recommended application at the ratio of 1:10 (silica gel to seed quantity) in an airtight jar. Seeds preserved with silica gel retained higher viability for seven years when stored at both 5 and –20°C. Silica gel is more useful at 5°C than at –20°C.

Onion seeds stored in desiccators under warm conditions preserved viability for nine years in spite of fluctuations in temperature. Likewise, Harrison opined that addition of silica gel to the storage container maintains seed viability under adverse storage conditions, while Horky believed that the presence of a desiccant did not affect seed quality and that germination rate remain unchanged. However, silica gel does improve the seed quality in the presence of fluctuating storage temperatures. Calcium oxide and dry rice are other desiccants used in onion storage. Dry rice preserves the onion seeds for three years at ambient temperatures.

*Seed packaging and sealed storage.* Packaging is essential to protect the seeds from high humidity by providing a suitable moisture barrier structure. Many packaging materials, such as kraft paper bags, muslin cloth bags, glass containers, polyethylene pouches, aluminum cans, and laminated aluminum foil pouches, are used for seed storage. The choice depends on the kind and quantity of seed material, temperature, and period of storage. Cloth bags and kraft paper bags are suitable for short periods of seed storage. For longer storage, moisture-proof containers and low temperatures are employed. Harrington reported that aluminum foil or tin cans are completely resistant to moisture penetration. Here, fluctuation of moisture occurs due to frequent opening and closing of the containers. Seeds of "Yellow Bermuda" onion remained viable for a longer period in sealed containers than in unsealed containers. In sealed storage, lowmoisture seeds preserve their viability , while high-moisture seeds (13 to 15 percent) could not maintain viability beyond 104 days when stored at 25°C.

Onion seeds stored in sealed jars at 15 to 20°C maintained storability longer than those stored in unsealed containers. Aluminum foil pouches are completely impermeable to moisture, and seeds stored in them retain initial viability for 22 months. It is recommended that onion seeds be stored in aluminum foil pouches at 5°C. These pouches are handy, occupy less space, and can be shaped to any size; they can be heat sealed and reused, and they do not release toxic compounds during storage.

Elemery preserved onion seeds at 32 percent RH, and viability was unaffected by packaging or temperatures, whereas germination was signifi- cantly affected at 75 percent RH when seeds were stored in jute-woven polyethylene or cloth bags, but not in polyethylene film covered with jute or cloth bags. Seed moisture content increases if seeds are stored in jute or cloth bags.

Onion seeds having 6.5 percent moisture content and packed in paper packets and in laminated aluminum foil pouches were stored under ambient conditions (16 to 35°C and 25 to 90 percent RH). Seed viability is well preserved in sealed storage and decreases rapidly under unsealed conditions . Laminated aluminum foil pouches and polyethylene bags are effective in preserving high seed viability for five years at 5 and –18°C. Viability was further extended to a period of 15 years by storage at –20°C. A well-dried onion seed can be stored for a longer period in sealed storage.

*Low-temperature storage.* Low-temperature storage is the most ideal method for onion seed storage. Seeds remain viable for fairly longer periods at low temperatures. The storage temperature is selected based on purpose and length of seed storage. High-quality seeds are preserved at subzero or ultralow temperatures for longer periods, while for bulk storage, seeds are stored above 0°C with a dehumidifier for shorter periods. Seeds are to be well dried for subzero temperature storage; otherwise, they will be affected by chilling injury. Moisture-proof containers such as metal cans, glass containers, and aluminum foil pouches are used for these purposes.

Seed deterioration is significantly reduced at lower temperatures. Cold storage of onion seeds is a very satisfactory method for conservation of high seed viability for longer periods. Onion seeds were successfully preserved at –20°C for 15 years without decline in viability and vigor, and there were no morphological variations in the seedlings that emerged from cold-stored seeds. According to Tronickova , seed germination and seedling emergence are improved following sealed storage at –14 to –20°C. Barton stored onion seeds at –4°C for 20 years. Likewise, Gane observed high viability at 0 and 10°C up to 60 percent RH, and –20°C is superior to 0 or 5°C. Rusev stored onion seeds in sealed glass containers at –1 to –2°C for three to four years.

*Modified-atmosphere storage.* Seed viability is preserved well in the absence of oxygen. Replacement of oxygen with other inert gases, such as carbon dioxide or nitrogen, is beneficial in maintaining high viability, especially at lower seed moisture levels. A high quantity of oxygen during storage affects seed viability. According to Roberts and Abdalla , the deleterious effects of oxygen are more pronounced at high levels of seed moisture. Harrington reported that onion seeds deteriorated faster when stored in oxygen, and slower when stored in nitrogen. Seeds stored in a nitrogen atmosphere showed high seed viability , and longevity improves with seed storage in nitrogen, more so than in carbon dioxide.

However, onion seed storage in carbon dioxide is more beneficial than storage in air. Harrison reported that carbon dioxide-stored seeds preserved high viability for six years. Gaseous storage is effective for short-term storage and beneficial in the absence of a cold-storage facility.

## Invigoration of Stored Seeds

*Seed deterioration.* Onion seeds are highly perishable , and reductions in germination and seedling vigor are the major criteria for the physiological manifestation of seed deterioration. Physical and chemical changes associated with deterioration are discoloration of seed coat, delayed germination, decreased tolerance to suboptimal environmental conditions during germination and storage conditions, reduced germinability and seedling growth, and increased number of abnormal seedlings.

Biochemical changes include an increase or decrease in enzyme activity, a decrease in oxygen uptake, an increase in leaching of organic and inorganic constituents from seeds, an increase in free fatty acids, a decrease in total soluble sugars, an increase in reducing sugars and a decrease in total soluble sugars, a decrease in proteins, and an increase in amino acids, and changes in carbohydrates, organic acids, and protein metabolism.

The exact cause of seed deterioration is not known. Banerjee observed that as long as the cells concerned with assimilation and transportation of food and the meristematic regions are alive, the seeds remain viable and are capable of germinating. Cell structure gets disturbed on deteriora tion, resulting in excessive leakage of electrolytes, soluble sugars, and potassium. In onion seeds, with increasing storage period, the palmitic and stearic acids increase and the oleic and linoleic acids decrease.

According to Suh, Lee, and Park , seed viability is negatively correlated with saturated fatty acid content and positively correlated with unsaturated fatty acid content. Seed vigor, an essential component for crop production and better storage, decreases on deterioration.

Normally, high seed vigor is estimated by percentage of germination. Kononkov, Kononkova, and Startsev developed a rapid technique for estimating seed quality. Seeds are kept in gauge bags placed in boiling water for 30 min. The number of seeds producing radicles is directly proportional to seed viability. Based on estimated storage conditions, Ellis and Roberts developed seed viability equations to predict seed storability. Stumpf, Peske, and Baudet showed that such equations are less accurate at higher levels of moisture and for longer storage periods. The possibility of using ATP (adenosine triphosphate) content as an indicator of onion seed quality was also studied.

A high positive correlation exists between ATP content and viability. ATP content is greater for seeds stored at 3°C than for those stored at 30°C. Likewise, dehydrogenase activity, which is directly related to seed viability, is greater in seeds stored at low temperatures than in those stored under ambient conditions. Exposing seeds to accelerated aging (42 or 45°C at 100 percent RH) for 72 h can help predict seed quality. Seed quality can be improved to a certain extent by imposing the following treatments.

*Seed priming.* Seeds have to be treated with osmotic solution to hasten the germination process. This gives rapid and uniform seedlings and delays aging. PEG is most effective in this regard. Primed seeds are dried to their original level of moisture and stored. This gives higher germination than unprimed seeds, a result attributed to the cellular repair process during the imbibition state. According to Basu , prestorage dry treatments, such as acetone permeation of antiaging chemicals and low concentrations of halogens (iodine, chlorine, and bromine) and alcohol (methanol, ethanol, and isopropanol), improve the storability of dry-stored seeds.

The intricacies involved in the invigoration of seeds are not clear. It appears that prevention of damaging oxidative reactions, such as free-radicalinduced peroxidation reactions involving unsaturated lipid molecules of lipoprotein biomembranes, and the repair of age-induced damage to vital bioorganelles by the cellular repair system are associated with seed invigoration.

Midstorage treatment with water or saline solution helps in retaining the original vigor and viability of seeds. Doijode and Raturi reported that hydration and dehydration improve the storage life and vigor of seeds. This improvement is associated with enhanced activity of dehydrogenase and peroxidase, with a simultaneous reduction in lipolytic activity and breakdown of fat contents. Onion seeds were soaked in water and other solvents, such as ethyl alcohol, acetone, xylene, dichloromethane, and petroleum ether. Seeds soaked in water followed by drying gave higher germination than other solvents. Highmoisture onion seeds (15 to 25 percent) were stored at high temperatures (45°C); the loss of viability was less for seeds having moisture of 20 to 25 percent than for those having 15 percent moisture. It is suggested that activation of the cellular repair process is due to imbibitions. Addition of monosodium hydrogen phosphate (10–4 M) and sodium sulfate (10–4 M) to water used for soaking gives minor additional advantages.

*Chemicals.* Chemical treatment of seeds is beneficial in delaying the aging process. Impregnation of acetone with vitamin E (20 units/milliliter [ml]) and butylated hydroxytolule (10–1 M) improves the storability of onion seeds. Natural and synthetic antioxidants can also be used, either alone or in combination with drying, to extend seed longevity. Six-month-old onion seeds treated with certain chemicals, such as EDTA (ethylenediaminetetraocetic acid) (0.1 M), potassium metabisulfite (KMS) (0.02 M), oxalic acid (0.01 M), ascorbic acid (0.02 M), PEG (0.5 M), and saturated glucose can be stored under ambient conditions (16 to 35°C and 25 to 90 percent RH). Seed germination and vigor improved with the treatment. These seeds were viable for a longer period than untreated seeds during ambient storage . These chemicals act as antioxidants or prevent free-radical formation. Leaching of toxic metabolites from seeds occurs during imbibitions. According to Dasgupta, Basu, and Basu , leaching of inhibitors is greater in the presence of dilute salt. EDTA prevents the formation of radical centers on unsaturated fatty acids and prevents the deterioration of seeds.

EDTA also extends seed storage life and vigor through the cellular repair system during the hydration phase.

*Growth regulators.* Seed quality improves with the application of growth regulators. Kinetin improved germination percentage and seedling vigor even in seeds stored under unfavorable storage conditions, while ABA lowered seed viability and vigor.

*Magnetic stimulus.* Electromagnetic treatment of seed causes morphological, physiological, and biochemical changes in seeds. Five-year-old onion seeds stored at 5°C can be exposed to electromagnetic fields (36 to 144 oersteds). Seed germination is improvedwith exposure to a 108 oersteds electromagnetic field for 30 min. This alsoimproved seedling length and dry weight, and leaching of metabolites decreasedas compared to untreated control. Magnetic stimulus promotes the biological mechanism by slowing down the process of seed deterioration and stimulates the seedling growth rate.

# Chapter 43

# LEEK: *ALLIUM AMELOPRASUM* VAR. *PORRUM* L.

## INTRODUCTION

Leek is a robust plant popular in the Middle East and Northern Europe. It does not form bulbs; instead, the long green leaves are cooked as a vegetable. The blanched stems and leaves are boiled or fried. It is a rich source of minerals such as phosphorus and iron.

### Origin and Distribution

Leek originated in the Mediterranean region. It is cultivated in Belgium, China, Denmark, France, Malaysia, the Netherlands, the Philippines, Spain, and the United Kingdom.

### Morphology

Leek belongs to family Alliaceae and has chromosome number of $2n = 32$. It is a monocotyledonous biennial plant. Leaves are flattened and variable in length. Inflorescence is a terminal umbel with many pinkish flowers. Fruit is a capsule. One thousand seeds weigh about 3.75 g.

## SEED STORAGE

Seeds are generally used in the propagation of leek. Also, seeds are used in crop improvement and in long-term conservation of genetic diversity. Seeds show orthodox storage behavior; they can tolerate desiccation and store well at chilling temperatures.

### Seed Collections

Leek is a cool-season crop. It grows well in low-temperature areas and at higher altitudes. It needs vernalization and long days for induction of flowering. It requires deep well-drained soil for optimum growth. The seeds are sown in the nursery, and then seedlings are transplanted in trenches in the field. Seed-to-seed or bulb-to-seed method is used to produce seeds. Leek is a cross-pollinated crop, and bees bring about pollination. An isolation distance of 1,000 m is maintained between two different varieties. Normal cultural practices, such as fertilizer, irrigation, and plant protection, are followed.

Off-type plants are removed. Mature heads are collected, dried, and threshed; seeds are separated, cleaned, and packed for storage.

### Seed Germination

Seed germination is rapid at low temperature. It is highest at 20°C and is affected at 30°C. Likewise, Corbineau, Picard, and Come reported that the optimum temperature for germination was between 15 and 20°C. Fresh seeds exhibit dormancy that persists for several months. Prechilling of seeds and exposure to an alternate temperature regime (20/30°C for 16/8 h) promotes germination.

### Storage Conditions

Leek seeds are short-lived under ambient conditions. Seed viability and vigor decrease with an increase in storage temperature and relative humidity. Seeds stored at 26.7°C gave 65 percent germination after six years of storage, and storage at 43.3°C further reduced seed viability. Litynski and Chudoba preserved high viability for two years at 2°C and 10 percent RH. Further, seeds with 7.7 to 13.8 percent mc stored at –20°C retain their original viability for five years. The stimulatory effect of priming persists after drying and subsequent storage with silica gel up to 15 months.

# ROOT CROPS

Chapter 44

# RADISH: *RAPHANUS SATIVUS* L.

## INTRODUCTION

Radish is a popular vegetable crop in Asia and Europe. It is mainly cultivated for the young tuberous root, which is used in salad. Tender and nonpithy roots are cooked as a vegetable and also made into pickles. The foliage is also used as a leafy vegetable in Asia. The immature pods and leaves of another species, *Raphanus caudatus,* are eaten raw and/or cooked as a vegetable. Radish increases the appetite and gives a cooling effect. Roots are rich in minerals and vitamins A and C. The pungency of radish root is due to a volatile substance called isothiocyanate. Radish has medicinal properties, especially in curing ailments of the liver and gall bladder, headache, diarrhea, piles, and jaundice.

### Origin and Distribution

Radish probably originated in Western Asia, including the greater portion of China, where wide genetic diversity exists. Later, its cultivation spread to temperate and tropical regions. It is grown in Brazil, Germany, Hungary, India, Indonesia, Italy, Japan, Malaysia, the Netherlands, the Philippines, Spain, the United Kingdom, and the United States.

### Morphology

Radish belongs to family Cruciferae and has chromosome number $2n = 18$. It is dicotyledonous annual or biennial herb. Its hypocotyls and swollen primary roots are used in salad. Root size, shape, and color vary in different cultivars. They are round, tapering, and cylindrical, either red or white, and flesh is white. Leaves are pinnate and partly divided. Flowers are small, white, or pink, self-incompatible, and pollinated by insects. Fruit is an indehiscent silique, with a long beak, and contains 6 to 12 seeds. Seeds are small, 3mmin diameter, yellow or brown, and globose. One thousand seeds weigh about 10 g.

## SEED STORAGE

Radish is raised commercially through seeds. Seeds show orthodox storage behavior and are capable of maintaining high viability, vigor, and genetic stability on long-term conservation. Radish seeds are fairly good storers under ambient conditions.

### Seed Collections

Radish is a cool-season crop. Asian types are suited to tropical conditions; European ones are less pungent and grow well in tropical plains, and produce seeds in hilly regions. Small radishes are adapted to temperate climates, and larger ones can tolerate fluctuations in temperature. Low temperature is congenial for development of root texture, flavor, and size. High temperature leads to premature bolting. Radish prefers loose, fertile, and well-drained soils. Seeds are produced either by root-to-seed or seed-toseed method. In root-to-seed method, the matured roots, called stecklings, are removed and replanted by cutting the tops. Seed size and weight decide seed quality. With increased seed size, the field emergence increases. Large seeds also give higher yield than smaller seeds. Proper cultural practices, such as adequate spacing, nutrition, irrigation, and plant protection measures, are to be followed for production of quality seeds. Radish is a cross-pollinated crop, and bees and other insects bring about pollination. Thus, a minimum isolation distance of 1,600 m is maintained between cultivars. The off-type roots are rogued based on shape and color. The crop is harvested when pods become brown and seeds near maturity. Dried pods are collected and threshed; seeds are removed, dried, cleaned, and packed for storage.

### Seed Germination

Fresh seeds show dormancy, and six weeks of after-ripening eliminates the dormancy. Further, germinating seeds using an alternate temperature regime of 10/20°C for 18/6 h breaks the dormancy. In radish, seed germination is epigeal and takes five days to complete. Fresh seeds absorb more water and germinate more slowly than older seeds. Seed germination improves by application of gibberellic acid (5 to 10 ppm) , naphthaleneacetic acid (10 to 20 ppm) , potassium nitrate , and acidulated water (pH 4.5) with nitric acid , and it is inhibited by exposing seeds to far-red light.

## Storage Conditions

Radish seeds retain their viability longer than other cruciferous seeds. The economic life of spring and winter radishes is 9 to 12 years. Seed germination decreased to 50 percent after 7 years of storage under ambient conditions. Prolonged seed storage increases the percentage of substandard produce and the risk of invasion by many diseases. Bold and well-matured seeds retain better viability than immature seeds.

Seed drying to lower moisture helps in better retention of seed viability. Well-dried seeds stored for 9 to 12 years at 10–1 to 10–2 millibar (mb) preserve high seed quality without loss of viability. Kretschmer and Waldhor stored lowmoisture seeds (4.2 percent) in plastic bags at –20°C for ten years. There was little change in moisture content, and initial germination (86 percent) remained unchanged. Likewise, seeds with 6 percent moisture packed in polyethylene bags and laminated aluminum foil pouches and stored at 5 and –20°C maintained viability for 15 years. The retention of viability during storage is better at –20°C than at 5°C. These stored seeds when sown in the field did not show any morphological abnormalities for root and shoot characteristics.

*Storage fungi.* Storage fungi such as *Alternaria alternata, Aspergillus flavus,* and *Fusarium moniliforme* associate with seeds during storage, especially under high levels of seed moisture or relative humidity. Seed treatment with captan, thiram, or carbendazim controls the fungi.

## Invigoration of Stored Seeds

Seed viability is lost gradually during storage. However, loss of vigor precedes the decrease in viability. Viability can be revived to a certain extent by exposing stored seeds to physical or chemical stimuli. Seed soaking in water followed by drying improves germination and seedling vigor. Soaking leaches out certain toxic substances and activates the enzymes involved in germination. Savino, Haigh, and Leo attributed this to the counteraction of free radicals that damage cellular components.

# Chapter 45

# CARROT: *DAUCUS CAROTA* L.

## INTRODUCTION

Carrot is a popular root crop in temperate, tropical, and subtropical regions. It is cultivated as a fresh produce item and also processed, canned, and dehydrated. Tender carrots are used in soups, stews, curries, and confectioneries and cooked as a vegetable. Carrot juice is a popular drink in de veloped countries. Carrots are rich in carbohydrates, fiber, minerals, especially iron, and vitamins, such as alpha- and beta-carotene. Carrot roots, particularly black cultivars, are fed to cattle. Carrot seed contains an essential oil that acts as a carminative and stimulant and has wide usage in the treatment of kidney ailments.

### Origin and Distribution

Carrot genetic diversity exists in Africa, Europe, and West Asia. It probably originated in Afghanistan. It is cultivated in almost all countries. Some leading producers are Argentina, Brazil, Canada, China, France, India, Indonesia, Italy, Malaysia, Mexico, Morocco, the Netherlands, Nigeria, the Philippines, Poland, Russia, the United Kingdom, and the United States.

### Morphology

Carrot belongs to family Umbelliferae and contains chromosome number 2n = 18. It is a dicotyledonous biennial plant with a short stem and welldeveloped roots. Its swollen roots, which are either short and blunt or long with a tapering cone, are edible. It is cultivated as an annual vegetable and as a biennial for seed production.

The stem elongates and gives inflorescence during the second year. The inflorescence is a compound umbel that is flat or round. Flowers open from the outer orbit and proceed toward the center. Flowers are perfect and white, with five stamens and an inferior ovary with two locules. Fruit is oblong-ovoid, ciliate, and indehiscent and contains a single seed, which contains essential oil. One thousand seeds weigh about 0.8 g.

## SEED STORAGE

Carrot is cultivated by seeds. Seeds are also used in crop improvement and in conserving genetic diversity through seed storage. Seeds show orthodox storage behavior; they can withstand desiccation and chilling during prolonged storage. Well-dried seeds retain viability for longer periods at low storage temperatures. Seeds with less than 10 percent moisture remain viable for six months under ambient conditions. Further, sealed storage preserves seeds for eight years.

### Seed Collections

Carrot requires a cooler climate for growth and development. The Asian types produce seeds in plains, while European types require vernalization of six to eight weeks at 8°C for induction of flowering. However, seeds are normally produced in hills, having higher altitude and a cooler climate. Car rot grows well in deep, sandy loam, well-drained soil. Seeds are produced either by seed-to-seed or root-to-seed method. The latter is a widely accepted practice, requiring the selection of good roots for planting. Plants are spaced properly, adequately fertilized, timely irrigated, and protected against pests and diseases. Planting full-size roots, with one-half or twothirds top, gives higher seed yield. A closer spacing of 45 × 20 cm also gave higher seed yield.

Flowering occurs at eight-day intervals in different orders of umbel (see Figure.1). Individual flowers are protoandrous, and insects, especially bees, bring about cross-pollination. An isolation distance of 1,600 m is maintained between varieties. Off-type plants showing variation for root and shoot characteristics are removed. The primary and secondary umbels are selected for seed purposes, as they give good-quality seeds. Seeds from primary umbels are heavier than those from secondary ones. The crop is harvested when the third-order umbels turn brown. Plants are cut at the base and stored for curing. Subsequently, seeds are separated, cleaned, dried, and packed for storage.

### Seed Germination

Fresh carrot seeds show a considerable amount of dormancy , which is attributed to underdeveloped embryos and immature seeds. Mature seeds also show dormancy .

According to Doust and Doust , an after-ripening period of three months eliminates seed dormancy.

*Figure.1. Carrot Flowering in Different Orders of Umbel*

Comparatively, wild relatives of carrots show more pronounced dormancy than cultivated ones. Seed germination is higher at a constant temperature from 15 to 25°C or at an alternate temperature of 20/30°C for 16/8 h, respectively. Seed germination is greater in vigorous seeds, which complete the germination process in a short time. Seeds with low vigor take longer for seedling emergence and exhibit greater variability in plant characteristics.

Likewise, seed germination is lower and delayed in immature as compared to mature seeds.

*Seed size.* Seed size also affects seed germination. Malik and Kanwar reported that field emergence is greater for large seeds. Likewise, large seeds (1.66 mm) have greater storability. The crop yield is also increased by 40 percent when large as compared to smaller seeds are sown.

*Chemicals.* Seed germination is improved by soaking seeds in naphthaleneacetic acid (100 ppm). Seed soaking in water exudates certain chemicals, including toxic ones. According to Chaturvedi and Murallia , these exudates contain a high amount of inhibitors that affect the germination process. The amount of exudates mainly composed of watersoluble sugars is negatively correlated with seed viability.

*Seed priming.* Seed priming promotes germination in fresh and old seeds. Seeds soaked in polyethylene glycol (6000) at 15°C for ten days resulted in early and higher germination.

## Storage Conditions

*Low-temperature storage.* Seeds attain full viability and vigor on maturity. Such seeds retain their viability for a longer period than immature seeds. Seed germination and seedling vigor decrease when seeds are stored at 20°C and 80 percent RH. Seeds lose viability rapidly at higher temperatures and relative humidity. The ideal storage conditions for longer storage life are temperature close to 0°C and 70 percent RH. Loss of seed viability is more rapid in dry seeds than in ultradry seeds at 20°C. Seeds with 2.0 to 3.7 percent mc (ultradry) and those with 5.5 to 6.8 percent mc (dry) retained viability for five years at –20°C, while seeds with 7 percent moisture exhibited 60 percent germination in unsealed, and 80 percent in sealed, conditions after two years of ambient storage. Seeds stored in moisture-proof containers at room temperature remained viable for 22 months. Seed stored in polyethylene bags maintained higher viability than those stored in paper bags. Seed viability was maintained for eight years at 5 and –20°C storage (Doijode, unpublished data). Likewise, Kretschmer and Waldhor reported that seeds stored in plastic bags maintained viability for 11 years at –20°C storage.

*Storage with desiccants.* Desiccants are effective in lowering seed moisture. Seeds stored with silica gel retained viability for 36 months under ambient conditions. Horky reported that desiccants are ineffective at a constant storage temperature of 0°C and they improve seed storage quality in the presence of fluctuating temperatures between 5 and 30°C.

*Modified-atmosphere storage.* A higher level of oxygen in the storage container affects seed viability. Several volatile compounds, such as methanol, ethanol, acetone, and acetyldehyde, are released from dry seeds during storage. The quantity of volatile compounds released increases with increasing storage period and temperature. Lowering the oxygen level and/or replacement with inert gases, such as nitrogen or carbon dioxide, maintains high germination during ambient storage.

*Storage of pelleted seeds.* Storage of pelleted seeds enhanced germination by 4 to 8 percent compared to nonpelleted seeds. Further, the yield from pelleted seeds is greater by 10 percent. *Storage fungi.* Storage fungi are largely associated with improper storage, especially when high seed moisture is retained during ambient storage. Fungi are effectively controlled by seed treatment with thiram, which also promotes seed germination. Elaine and Shiel suggested coating seeds with resins and incorporating fungicides during storage to protect seeds from fungal attack.

## Invigoration of Stored Seeds

Soaking stored carrot seeds in water or in dilute chemicals, such as sodium thiosulfate (10–5 M) or disodium phosphate (10–4 M), for 2 h followed by drying reduces physiological deterioration during subsequent storage. Even hydration by saturated atmosphere for 24 to 48 h followed by drying is effective in reducing loss of seed viability. Midstorage hydration-dehydration is effective in improving seed viability and vigor.

# Chapter 46

# BEETROOT: *BETA VULGARIS* L.

## INTRODUCTION

Beetroot is an important salad crop in temperate regions. It is also known as garden beet. Here, the edible portion includes the swollen hypocotyl and a smaller part of the taproot. The foliage is also used as greens and cooked as a vegetable. Apart from its usage in salad, beetroot is boiled, pickled, or cooked as a vegetable. It is a good source of carbohydrates, calcium, phosphorus, potassium, and vitamin C. Its foliage is rich in iron and vitamin A.

### Origin and Distribution

Beetroot originated in Asia Minor, the Mediterranean region, and southern Europe. It is a popular vegetable crop in Europe, the Middle East, and North America. It is cultivated in Canada, China, Denmark, Germany, India, Indonesia, Italy, Malaysia, the Netherlands, the Philippines, the United Kingdom, and the United States.

### Morphology

Beetroot is a biennial glabrous herb that is cultivated as an annual for vegetable purposes. It belongs to family Chenopodiaceae and has diploid chromosome number 2n = 18. The food is stored in the root, which is flat, conical, or tapering; red; and swollen. The red pigment is due to the presence of betacyanins. Foliage is dark green or red and ovate, forming a rosette. Inflorescence is a three- to four-flowered cyme; it grows up to 1 m in height.

Flower is small, with a green calyx, no corolla, and five stamens, with one ovule per ovary. Fruit is an aggregate formed by the partial fusion of two or more fruits. In this, the calyx continues to grow after fertilization; becomes hard, uneven, and corky; and encloses the seed. It is referred to as beet seed or glomerule and normally contains two to six seeds. True seeds are small, brown, 3mmlong, and kidney shaped. One thousand seeds weigh about 17 g.

## SEED STORAGE

Beetroot seeds show orthodox storage behavior. The retention of high seed quality, viability, and vigor is the desired end in seed storage. Seeds with high initial vigor maintain viability for a longer period. Seeds are viable for two to six years under normal storage conditions.

### Seed Collections

Beetroot is a cool-season crop, requiring a cool and humid climate for optimum growth. It needs deep, fertile, well-drained, sandy loam soils. Heavy soils are unsuitable for cultivation, and it is sensitive to waterlogging. Beetroot is propagated by seeds. Plants require vernalization for induction of flowering, and an altitude over 1,000 m is congenial for seed production. Seeds are commonly produced by the root-to-seed method, which allows root inspection during transplanting. Sometimes the seed-to seed method is also used. The first healthy and true-to-type stecklings are selected and stored for a short period before replanting. Proper fertilizer dosage and irrigation, especially during the dry period, are applied to the seed plots. Plants are protected from pests and diseases. Beetroot is crosspollinated, predominantly by wind, so an isolation distance of 1,600 m is maintained between cultivars for maintenance of genetic purity. Off-type plants are removed based on foliage and root characteristics. Fruits become brown on maturity (see Figure.2). Plants are cut when the fruits on lower branches are mature, then they are heaped and stored for a short period. Later, the dried fruits are threshed; seeds are separated, cleaned, dried, and suitably packed for storage.

### Seed Germination

Seed germination is slow and erratic and takes longer to complete. Fruit surface contains an inhibitor that affects the germination. Simple washing or soaking in water or in plant extracts of certain seaweeds, which appear to possess cytokinin-like substances, improves seed germination. Khan et al. reported that osmoconditioning of seeds with 1.2 MPa of PEG (6000) followed by a two-minute water rinse is effective in improving germination.

*Figure.2. Beetroot Fruits on Maturity*

## Storage Conditions

Beetroot seeds remain viable for a fairly long period under room temperatures. However, they deteriorate rapidly at higher storage temperatures and humidity. Lowering of seed moisture is beneficial for prolonging storage life and maintaining high seed quality. Seed moisture is reduced by 1 percent through the addition of sodium sulfate at the rate of 1.5 percent of seed weight basis. Seed storage at lower temperature reduces the metabolism and slows the process of seed deterioration. Seeds stored at 10°C and 50 percent RH maintained high seed viability for 13 years.

# Chapter 47

# TURNIP: *BRASSICA RAPA* L.

## INTRODUCTION

Turnip is an important cruciferous vegetable crop in temperate regions that is, of late, becoming popular in tropical regions as well. It is cultivated for its fleshy roots and foliage. Tender roots are used in salad and are cooked as a vegetable. It is a good source of iron and vitamins A and C.

### Origin and Distribution

Turnip originated in the Mediterranean region and is cultivated throughout the world. It is grown in Afghanistan, Canada, Denmark, France, Germany, India, Italy, Japan, Malaysia, the Philippines, Russia, the United Kingdom, and the United States.

### Morphology

Turnip belongs to the Cruciferae family, and its diploid chromosome number is 2n = 20. It is a biennial dicotyledonous herb that is cultivated as an annual for vegetable purposes. Food is stored in the taproot, which is flat, globose, or a long globe. Its color above ground is red, purple, white, or yellow, and the underground portion is white or yellow. Foliage is light green and hairy, with lobed leaves. Flower is perfect and bright yellow, with five sepals, five petals, five stamens, and a three-celled superior ovary. Fruit is a dehiscent silique; seeds are round; black, red, or brown; and 1.5 to 2 mm in diameter. One thousand seeds weigh roughly 4.3 g.

## SEED STORAGE

Seeds are used in the commercial production of turnip. They are also employed in evolving new genotypes and in maintaining genetic diversity for longer periods. Seeds show orthodox storage behavior and remain viable for longer periods under dry, cool storage conditions.

### Seed Collections

Turnip is a hardy cool-season crop that can withstand frost. It grows well in deep, fertile, well-drained, sandy loam soils. Seeds are produced either by seed-to-seed or root-to-seed method. The latter is commonly used and provides easy selection of desirable types of roots for high seed quality.

Healthy true-to-type roots are transplanted. The inflorescence is a terminal raceme. It is a cross-pollinated crop, and honeybees are the main pollinators. It is easily crossable with Chinese cabbage, radish, and mustard. Thus, an isolation distance of 1,600 m is provided to safeguard against genetic contamination. Roguing is done at different times to remove off-type plants. Normal and timely cultural practices, such as fertilizer application, irrigation, and suitable plant protection measures, are followed during seed production. Pods turn yellow on maturity, and the whole flowering stalk is cut, heaped, and allowed to cure for four to five days. Pods are threshed, and seeds are separated, cleaned, dried, and packed for storage.

### Seed Germination

Seed germination is epigeal and higher at a constant temperature of 20°C or at alternate temperatures of 20/30°C for 16/8 h. Seed size varies from 1 to 1.7 mm; large to medium-sized seeds give early and higher germination than smaller seeds.

### Storage Conditions

Turnip seeds store well under ambient conditions. Higher storage temperatures and relative humidity are injurious to seed longevity. Lazukov reported that seeds stored at 18 to 22°C and 60 to 80 percent RH remained viable for 12 to 13 years. Economic seed longevity is about two to three times less than biological longevity. Seed moisture is lowered and kept at a safe limit through the use of chemical desiccants. According to Nakamura , calcium chloride is a satisfactory desiccant for longer seed storage. Another desiccant, sodium sulfate, applied at the rate of 1.5 percent of seed weight reduces the seed moisture by 1 percent and keeps seed viable for a longer period. It also protects seeds from pathogens. Resins are used in storage to prevent premature germination and to facilitate storage by extending storage life.

# COLE CROPS

# Chapter 48

# CABBAGE: *BRASSICA OLERACEA* VAR. *CAPITATA* L.

## INTRODUCTION

Cabbage is an important winter vegetable crop widely cultivated in Europe and North America. It is also popular in the tropics and subtropics. It is cultivated for its tender, modified edible leaves, which form a compact structure called a head. Leaves are used for salad purposes, processed for pickles, sauerkraut, and cooked as a vegetable. Cabbage is a rich source of amino acids, vitamins A and C, and minerals such as phosphorus, calcium, potassium, sodium, and iron.

### Origin and Distribution

Cabbage originated in the Mediterranean region. It is grown in China, Colombia, Czechoslovakia, Denmark, Egypt, France, Germany, India, Indonesia, Italy, Japan, Korea, Malaysia, the Netherlands, the Philippines, Poland, Russia, South Africa, Spain, Taiwan, Ukraine, the United Kingdom, and the United States.

### Morphology

Cabbage belongs to family Cruciferae and has a chromosome number of 2n = 18. It is a biennial but is cultivated as an annual. The stem is short and encircled by compact foliage. Cabbage heads are round, flat, or pointed. Leaves are red, green, smooth, or wrinkled (see Figure.1). Flowering is induced on vernalization. Flowers are small, hermaphrodite, and yellow, with four sepals and petals, six stamens, and an ovary with two locules. Fruit is a silique that dehisces on maturity and contains 12 to 20 seeds. Seeds are small, smooth, brown, and globular. One thousand seeds weigh about 3.3 g.

*Figure.1. A Portion of a Cabbage Seed Plot*

## SEED STORAGE

Cabbage seeds are short-lived under ambient conditions. They show orthodox storage behavior and can withstand drying as well as chilling temperatures on storage. Seeds are stored for short periods for growing of crops, and for longer periods for the conservation of genetic diversity. A high plant population is a prerequisite for successful crop production, and achieving the required plant density depends on seed quality. Thus, to meet the requirement, suitable storage practices have to be followed for maintaining high viability and vigor during storage.

### Seed Collections

Cabbage is a cool-season crop that withstands frost. Hilly regions are generally well suited for seed production because chilling temperatures are prevalent. Normally, temperatures less than 10°C for five or six weeks are required for the induction of flowering. Cabbage grows well in deep, fertile, well-drained soils. Seeds are produced by head-to-seed or seed-toseed method. In the former method, the crop is raised for head purposes, and well-developed healthy and mature heads are selected for seed purposes and stored for a short period before replanting. Later, these are planted in the field by making a crosscut on top of the heads, without causing injury to the growing points. This facilitates early emergence of flower stalks. The crop is adequately fertilized, irrigated during dry periods, and protected against pests and diseases. In the seed-to-seed method, selected heads are left in the field and crosscuts are made to facilitate flower emergence. Cabbage is a cross-pollinated crop, and bees and flies mainly bring about pollination.

Crops are separated by a distance of 1,600 m from another varietal block. Pods turn yellow on maturity, and at this stage, they are collected, dried, and threshed; seeds are separated, cleaned, and suitably packed for storage.

## Seed Germination

Seed germination is epigeal and higher at a constant temperature of 20°C. Seed germination in dormant seeds can be improved by exposing to alternate temperatures of 20/30°C for 16/8 h, respectively. Seed color changes to dark brown at maturity but lightens on seed aging. The germination percentage is higher in darker seeds than in lighter ones. According to Liou and colleagues , low germination and emergence rates are associated with low-vigor cabbage seeds, which are improved by soaking seeds in water followed by osmopriming with polyethylene glycol (PEG).

*Storage of germinated seeds.* Germinated cabbage seeds are viable for a short period. Storage life could be extended up to five days if stored in aluminum foil pouches at 5°C. Longer storage reduces seedling emergence. The decrease in vigor is attributed to degradation of structure and enzymic activity. According to Finch-Savage and McQuistan , germinated seeds can be stored for one week at −20°C when cooled at the rate of 2 to 6°C per min.

## Storage Conditions

Initial seed vigor, seed moisture, storage temperature, relative humidity, and oxygen content regulate seed storability. Higher seed moisture and higher temperatures promote germination and deterioration. Elaine and Shiel suggested the use of resins for preventing premature seed germination. Seed coating with resin gives better storage but does not affect the germination process.

*Seed moisture.* High seed moisture reduces the storage life of seeds. Mackay and Flood reported that seeds with low initial germination and high moisture content deteriorated faster at 11 to 19°C and 58 to 83 percent RH. Seeds having 3 percent moisture retained their viability for 22 years at −10°C. Seeds readily absorb moisture from the atmosphere; thus, they must be packed suitably to protect against high humidity.

Various kinds of packaging, such as glass jars, polyethylene bags, and aluminum foil pouches or cans, are used for seed storage. Normally, lowmoisture seeds are sealed in packages before storage. Zink and Camargo reported that seeds with high moisture content packed in plastic bags deteriorated faster and lost viability after 16 months of storage, whereas low-moisture seeds maintained high viability for 54 months. Seeds attain moisture equilibrium earlier in hessian (burlap) or cloth bags than in polyethylene bags. Lowig noted that polyvinylidene chloride (PVDC) strengthened paper bags are effective moisture barriers, preserving seed viability for seven years.

*Seed storage with desiccants.* Seed moisture removal by desiccants is faster than air drying. Desiccants are effective in maintaining high seed viability under fluctuating temperatures. Erofeev and Zelenin recommended that cabbage seeds be mixed with sodium sulfate (1.5 percent) on a seed weight basis for a 1 percent reduction in moisture. With this, high seed viability was preserved for two years, and fungal microflora were reduced on seeds. Further, seed longevity was extended to eight years for desiccant stored seeds as compared to nine months for seeds in open storage.

*Temperature.* Cabbage seeds lose their viability rapidly at higher storage temperatures. They lost viability completely by four months of storage at 40°C and 77 percent RH. Horky reported that seed deterioration is greater at variable ambient temperatures, but seeds preserve high viability for 86 months at 0°C. Similarly, Lazukov preserved cabbage seed viability for 7 to 15 years at 18 to 22°C and 60 to 80 percent RH. Low-temperature cabinets, such as a freezer, are effective in preserving seed viability. Accordingly, seeds stored in airtight jars at –20°C preserved their viability for 23 years.

*Air.* Higher oxygen levels in storage containers increase the respiration rate, resulting in quick deterioration. According to Dressler , cabbage seeds stored in a vacuum at 10–1 to 10–2 mb maintain viability for 9 to 12 years. Also, high levels of carbon dioxide inhibit the germination process,and help in preserving seed viability (Thornton.

*Storage fungi.* Cabbage seeds associate with several storage fungi. Seed viability is affected by the presence of *Alternaria, Aspergillus, Fusarium, Rhizopus,* and *Sclerotium* spp. These can be effectively controlled through seed treatment with captan or carbendazim. Miller and Linn reported that cabbage seeds treated with thiram could be stored for 12 months at 2 to 32°C, with no loss of viability, and treated seeds germinated better than untreated ones.

# Chapter 49

# CAULIFLOWER: *BRASSICA OLERACEA* VAR. *BOTRYTIS* L.

## INTRODUCTION

Cauliflower is an important cole crop cultivated for its edible curd. The curd is made of highly metamorphosed abortive flowers on thick hypertrophied branches. It is cooked as a vegetable and eaten raw. The green foliage serves as good cattle feed. Cauliflower is rich in minerals such as potassium, sodium, iron, phosphorus, calcium, and magnesium. It grows well in temperate regions.

### Origin and Distribution

Cauliflower, which evolved from wild cabbage, is native to the Mediterranean region. It is largely cultivated in Australia, Canada, Caribbean islands, China, Egypt, France, Germany, India, Italy, Malaysia, the Netherlands, the Philippines, the United Kingdom, and the United States.

### Morphology

Cauliflower belongs to family Cruciferae and has chromosome number 2n = 18. It is a dicotyledonous biennial plant cultivated as an annual. It has a short, stout stem with long, dark green foliage that encloses the curd, protecting it from sun. The inflorescence is dense and terminal, with a large number of flowers. Flowers are small and have four sepals and petals, six stamens, and a bicarpellary gynoecium. Self-incompatibility exists in certain cultivars. Fruit is a silique about 8 cm long. Seeds are round and dark brown. One thousand seeds weigh about 2.8 g.

## SEED STORAGE

Seeds are widely used in cultivation, and as such, farmers prefer to preserve seeds for shorter periods, such as until next growing season. Seeds are also used in evolving new cultivars and for long-term conservation of genetic diversity. Cauliflower seeds show orthodox storage behavior and withstand desiccation and chilling temperatures during storage. Seeds are short-lived under ambient conditions.

Seed longevity is primarily controlled by genetic factors and further regulated by storage conditions. Doijode reported that seed storability varies in different cultivars. On deterioration, poor-storer cultivars showed greater leakage of electrolytes and soluble sugars as compared to goodstorer ones. Good-storer cultivars preserve viability for relatively longer pe riods under ambient conditions and do not require any special packaging or conditions for short-term storage.

### Seed Collections

Cauliflower is a cool-season crop and is less tolerant to heat and dry conditions. It grows well in cool, moist, fertile, well-drained soils. Seeds are sown in the nursery, and young seedlings are later transplanted in the field.

Normal cultural practices, such as timely fertilizer, irrigation, and plant protection measures, are followed. Seeds are produced either by growing in situ or by transplanting selected curds. Normally, plants with good curds are allowed in the field, and later they produce flowers and seeds. These selected plants are scattered all over the field, occupying a large cultivable area. To overcome this, selected plants are transplanted carefully to a seed production block. Curds are scooped during transplanting to facilitate flowering.

Cauliflower is highly cross-pollinated, and honeybees and bumblebees are the major pollinators. Thus, an isolation distance of 1,000 m is maintained between varieties. Off-type plants are removed. Pod ripening is nonuniform, and pods on lower branches mature first. These are pooled, cured for a few days, and threshed; seeds are separated, cleaned, dried, and suitably packed for storage.

### Seed Germination

Seed germination is epigeal and takes five to seven days to complete. Further, it is enhanced by exposing seeds to 20/30°C for 16/8 h and to light. Verma and Pujari noted that short-duration exposure of seeds to light increased the germination percentage.

## Storage Conditions

Cauliflower seeds are viable for a short period under room temperatures. They remain viable for nine months under open storage. Verma and colleagues reported that well-dried seeds (5 percent mc) packed in aluminum foil pouches could retain high viability under ambient conditions. Likewise, Doijode obtained a high germination percentage and high seedling vigor in seeds of good-storer cultivars after five years of storage in ambient conditions. Tronickova recorded greater loss of viability at 15 to 20°C, while maintaining high viability at −20°C for five years. Seeds removed from −20°C chambers further maintained high viability for three months under normal temperatures. Seeds stored in desiccators remained viable for eight years under ambient conditions , whereas those stored in plastic containers at −20°C, for six years.

*Storage fungi.* Storage fungi affect seed health. These fungi grow rapidly in humid conditions. Fungi such as *Alternaria alternata, Alternaria brassicicola, Aspergillus* spp., *Penicillium* spp., and *Rhizopus stolonifera* associate with seeds. These fungi cause discoloration and shriveling, increase storage temperature, and reduce seed viability. Maude and Humpherson-Jones observed 11 genera on cauliflower seeds, namely, *Alternaria, Aspergillus, Chaetomium, Cladosporium, Curvularia, Drechslera, Fusarium, Gloesporium, Phoma, Rhizopus,* and *Stemphylium.* Seed treatment with fungicides such as thiram or carbendazim is effective in controlling storage pathogens.

# Chapter 50

# KOHLRABI: *BRASSICA OLERACEA* VAR. *GONGYLODES* L.

## INTRODUCTION

Kohlrabi is also known as knol khol. It is widely cultivated in temperate regions for its swollen edible stem, and its cultivation is expanding to the tropics and subtropics. Kohlrabi is cooked as a vegetable. It contains high amounts of calcium, magnesium, phosphorus, sodium, sulfur, potassium, and vitamins A and C.

### Origin and Distribution

Kohlrabi originated in the Mediterranean region. It is largely cultivated in the Caribbean islands, Denmark, France, Germany, India, Indonesia, Italy, Malaysia, the Netherlands, the Philippines, and the United Kingdom.

### Morphology

Kohlrabi is a small, dicotyledonous, biennial herb that is cultivated as an annual. Its stem is small and stout and enlarges just above the ground in a sphere shape. The swollen stem is 5 to 10 cm in diameter and white, green, or purple. Flowers are light yellow in color. Fruit is a silique that dehisces on maturity. Seeds are round, 1.5 to 2.0 mm in diameter, and brown. One thousand seeds weigh about 3.2 g.

## SEED STORAGE

Kohlrabi is propagated by seeds. Seeds are also used in hybridization programs for evolving new cultivars and in long-term conservation of genetic variability. Kohlrabi seeds show orthodox storage behavior. Drying and cold storage do not affect seed viability. Kohlrabi seeds are short-lived under ambient conditions. They require suitable storage conditions for longer maintenance of high-quality seeds.

### Seed Collections

Kohlrabi is a cool-season crop. It requires low temperature (10°C) for a week to induce flowering. According to Marrewijk , cold treatment for about 8 to 12 weeks followed by 2 weeks at 10°C initiates flowering. Plants are earthen up or they are protected from frost by covering their swollen stems with soil, lifted and stored separately during frost. Normally, the seed-to-seed method produces seeds. Kohlrabi requires deep, fertile, and well-drained soils. Seeds are sown in the nursery, and seedlings are subsequently transplanted to the main field. Plants are fertilized properly, irrigated, and protected against pests and diseases. Uniform and true-to-type plants are selected for seed purposes, and off-types are eliminated. Kohlrabi is cross-pollinated by bees; thus, an isolation distance of 1,600 m is provided to maintain genetic purity. Mature pods are harvested, stored for a few days, and threshed; seeds are extracted, cleaned, dried, and packed for storage.

### Seed Germination

Seed germination is epigeal and completes in a week. It is rapid and highest at a constant temperature of 30°C.

### Storage Conditions

Kohlrabi seeds lose viability and vigor rapidly under ambient conditions. Tronickova reported that seeds stored at 15 to 20°C lost viability rapidly. In open conditions, seeds remained viable for nine months, and longevity was extended to eight years by preserving seeds in desiccators. Seeds stored at subzero temperatures (–20°C) remained viable for six years. Further, seeds maintained viability for three months at normal temperatures after storage at –20°C, with no reduction in germination or seedling vigor.

# POD VEGETABLES

# Chapter 51

# FRENCH BEAN: *PHASEOLUS VULGARIS* L.

## INTRODUCTION

French bean is widely cultivated in temperate, subtropical, and tropical regions. It is also known as snap bean, kidney bean, common bean, green bean, haricot bean, bush bean, navy bean, pole bean, string bean, and frijol. It is cultivated for its edible tender pods and dry seeds. Fleshy pods and immature seeds are cooked as a vegetable, processed for canning, or frozen.

Dry seeds are commonly used as a grain legume. French bean is rich in proteins, carbohydrates, potassium, and iron contents.

### Origin and Distribution

French bean originated in Central America. Large genetic diversity among primitive cultivars, landraces, and wild relatives exists in Guatemala, Honduras, and southern Mexico. It is largely cultivated in Brazil, Chile, China, Egypt, England, Ethiopia, France, India, Italy, Kenya, Peru, the Philippines, Spain, Tanzania, Turkey, Uganda, the United States, and Venezuela.

### Morphology

French bean belongs to the Leguminosae family and has chromosome number 2n = 22. It is a dicotyledonous, bushy, or twining annual vine. Leaves are dark green, alternate, and trifoliate. Flowers are self-fertilized, axillary, with one raceme, and white, pink, or yellow. Pods are narrow, long, fibrous, fleshy, or stringless and contain about 4 to 12 seeds. Seeds are globular to kidney shaped, and color varies from white, yellow, greenish, pink, red, purple, and brown to black. One thousand seeds weigh roughly about 250 to 600 g.

## SEED STORAGE

French bean is raised through seeds. Seeds are commonly used in hybridization programs for crop improvement and in the gene bank for maintenance of genetic variability for longer periods. French bean seeds show orthodox storage behavior. Seed viability is unaffected by decrease in seed moisture content and storage temperature. Seeds survive for a short period under ambient conditions. Further, seed quality during storage will be affected due to injury by mechanical threshing. Borges, Moraes, and Vieira reported that about 30 percent of seeds were viable after 30 days, and 18 percent after 90 days, of storage under ambient conditions. James, Bass, and Clark opined that year of seed productivity influences seed longevity, especially under poor storage conditions, but not under controlled ones. However, seeds remain viable for four years at 21.1°C and 50 percent RH. Well-stored seeds germinate and use their food reserves quickly, while poorly stored seeds use only one-fourth of their reserves before cotyledons drop. Such seedlings grow slowly and mature late, and they also yield less, by 27 to 30 percent.

Seed storability varies in different genotypes to a limited extent. Moreno- Martinez and colleagues reported that seed longevity varies from 60 to 180 days in different cultivars when stored at 25°C and 75 to 85 percent RH. Cultivars with white seeds lose seed viability more quickly than nonwhite- seeded cultivars. According to Hernandez- Livera and colleagues , seeds of cultivars Pinto, Texcoco, and Negro Puebla showed more vigor and viability on storage and were considered to have higher storage potential. These may be used as parents for improving seed longevity in offspring.

Vieira reported that aging of seeds was accompanied by low germination percentage and an increase in the percentage of abnormal seedlings. Such aged seeds give low yield. As the seed germination decreases, the leakage of amino acids, electrolytes, and soluble sugars increases , and seed starch content decreases, while proteins are unaffected. Free fatty acids and peroxidase increase with an increase in acidity value during storage.

## Seed Collections

French bean grows well in humid and cooler climatic conditions. High seed quality is obtained from seeds produced in cooler conditions. It grows well in moderate rainfall areas. It is propagated by seeds, which are sown directly in the field in rows. Seeds germinate better when the soil temperature is 30°C, while temperatures greater than 35°C are deleterious to germination. Soils rich in nutrients with good drainage are preferred for seed pro duction. French bean is sensitive to soil moisture stress, and most sensitive to salinity. It is a self-fertilized plant; however, an isolation distance of 50 m is maintained between varieties for genetic purity.

High temperature during flowering affects fruit set. Likewise, excess rainfall damages the crop and encourages pathogens. Plants are suitably protected by timely spraying of plant protection chemicals. Crops are adequately fertilized and regularly watered. Plants are checked during the flowering and fruiting stages, and off types are removed. Pods turn to yellow on maturity, and delay in harvesting shatters the seeds. Dried pods are collected and threshed, and the seeds are extracted, cleaned, dried, and packed suitably for storage.

*Seed maturity.* French bean seeds grow rapidly after fertilization. There is a continuous accumulation of dry matter in the seeds. On physiological maturity, seeds attain maximum weight and further inflow of dry matter ceases. Simultaneously, pods change color, becoming yellow, which is a useful sign in identifying seed maturity. According to Sanhewe and Ellis , longevity of developing seeds continues to increase after mass maturity. Maximum seed quality is attained at the end of the seed-filling phase in bean. The slow desiccation occurring during natural maturation is beneficial for development of high seed quality. Seeds of early harvested pods (60 days after sowing) gave 25 percent germination, and germination increased to 99 percent in seeds extracted from fruit pods 75 to 93 days old.

## Seed Germination

Seed germination is rapid and epigeal, and it takes about five days to complete. Germination is higher at a constant temperature of 30°C or at alternate temperatures of 20/30°C for 16/8 h. Carvalho and Oliveira reported that seeds sown with the hilum in the down position germinated earlier and gave the highest germination percentage. However, Bowers and Hyden recommended that laying seeds in the flat position is better for higher germination. Bold seeds with higher specific gravity produce healthy and vigorous seedlings and outyielded seeds of lower specific gravity. Similarly, higher seed germination and more vigorous seedlings were obtained from larger and heavier seeds. A direct positive correlation is found to exist between seed size and seedling size in French bean. Certain physical stimuli also promote germination. Seed exposure to 40 MHZ radio frequency electric fields enhances seed germination percentage by reducing hard seed contents.

## Storage Conditions

*Seed moisture.* Seed moisture plays a vital role in seed deterioration. Higher seed moisture promotes rapid metabolism and invasion of pathogens, thereby affecting seed quality. Lowering of seed moisture increases storage life. Zink, Almedia, and Lago reported that moisture content of 6 to 7 percent maintains high viability for 36 months, compared to less than nine months in seeds having 14.2 percent moisture. In another study, seeds having moisture of 9 to 10 percent stored at 11 to 20°C preserved high seed viability for 16 years. Seeds lose viability rapidly, especially at high moisture levels in sealed storage. Seeds with 12.4 to 20.1 percent mc maintain viability for one year when stored in paper bags. When dried carefully, seeds stored at 10–1 to 10–2 mb retained viability for 9 to 12 years without loss of vigor.

*Seed packaging.* Seeds are hygroscopic, and they exchange moisture rapidly with the atmosphere, especially at higher temperatures. To keep the moisture relatively near desired levels in seeds, suitable packaging is used.

According to Monteiro and Silveira , foam boxes, polyethylene boxes, and tin cans are ideal for seed storage, and seeds remain viable for 24 months. Seeds stored in sealed tin cans remained viable for two years at 15°C and 50 percent RH. Seeds stored in plastic bags at –20°C retain viability for 11 years. Seeds having 6.5 percent moisture packed in polyethylene bags and laminated aluminum foil pouches retained viability greater than 85 percent at 5 and –20°C after 15 years of storage, without loss of seedling vigor. In humid areas, it is advisable to pack seeds under sealed conditions, even if storage temperature is as low as 5°C.

*Temperature.* Low-temperature storage is beneficial for extending the storage life of seeds. Seeds stored at ambient temperatures gave lower germination percentage and lower seedling vigor than those stored in controlled environments. According to Toole and Toole , French bean seeds are short-lived under ambient temperatures.

Seeds remained viable for 24 months under ambient conditions and, thereafter, viability decreased until it was lost completely by the forty-eighth month of storage. Seedling vigor decreases with an increase in storage period. Seeds stored at higher temperature exhibit poor seed quality and accelerated darkening of seed coat.

Hughes and Sandsted noted that seeds stored at 24°C became dark and had poor germination and increased fatty acid content after one year of storage, while seeds stored at 1°C retained their original color and initial viability.

In another study, seeds darkened to a greater extent at 25°C than at 7°C. Low tannin contents are associated with dark seeds. Seeds stored at room temperature died after two to three years of storage, but those stored at 12°C and 45 percent RH retained high viability for nine years. With a rise in temperature and relative humidity, the storage period of seeds decreases. Seeds remain viable for 45 months at 17°C and 35 percent RH, for 30 months at 19°C and 57 percent RH, and for 27 months at 21°C and 57 percent RH. Seeds stored under favorable conditions (19°C and 57 percent RH) gave 23 percent higher germination and higher yield than seeds stored under unfavorable conditions.

Ultralow temperatures are also helpful for maintaining high viability and vigor. However, the utility of this method depends on the regular availability of liquid nitrogen, the size and quantity of seeds to be stored, and cost. Stanwood and Roos stored French bean seeds in liquid nitrogen (–196°C) for 180 days, with no reduction in germination percentage or vigor.

*Storage fungi.* High seed moisture or high relative humidity is congenial for the growth and development of storage fungi. Seeds with higher moisture (15 percent) are rapidly invaded by fungi, and their viability is affected. Infestation with *Aspergillus* spp. is more harmful than with *Geotrichum* sp. Seeds with lower moisture content retain their color and viability. Low moisture level in seeds is maintained by using moisture-proof containers, such as tin boxes, which check the growth of microorganisms. Seed deterioration occurs due to toxins produced by fungi that damage the cellular structure. Use of fungicides such as captan, thiram, and carbendazim reduce the damage caused by fungi.

# Chapter 52

# GARDEN PEA: *PISUM SATIVUM* L.

## INTRODUCTION

Garden pea is also called pea, sugar pea, or Chinese pea. It is widely cultivated in temperate regions for its edible fresh green seeds and pods. The fresh seeds are canned, frozen, or dehydrated and cooked as a vegetable. Dried peas are used as grain legumes. Garden pea is rich in proteins and carbohydrates and contains calcium, phosphorus, and vitamins A and C.

### Origin and Distribution

Garden pea originated in Central or Southeast Asia. It is a cool-season crop cultivated in Australia, Belgium, China, Egypt, Ethiopia, France, Hungary, India, Italy, Luxembourg, Myanmar, the Philippines, Russia, Turkey, the United Kingdom, and the United States.

### Morphology

Garden pea belongs to the Leguminosae family and has diploid chromosome number $2n = 14$. It is an annual dicotyledonous climbing herb with a hollow stem. Leaves are alternate and pinnate with terminal tendrils. Flowers are papilionaceous, hermaphrodite, and axillary, with two or three racemes; have five sepals and five petals; and are white, pink, or purple. Fruit is a pod that varies in size, is straight or curved, and contains two to ten seeds. Seeds are globose, wrinkled or smooth, and green, gray, or brownish. Wrinkled seeds are sweet. One thousand seeds weigh roughly 150 to 330 g.

## SEED STORAGE

Garden pea is propagated by seeds. Seeds are also used in crop improvement and for long-term conservation of genetic diversity. Seeds show orthodox storage behavior. Those with low moisture at low temperature remain viable for longer periods, and when stored under ambient conditions with moisture seldom exceeding 10 percent, they are viable for 18 months. Optimum moisture content for storage decreases as the storage temperature decreases.

Seed longevity is primarily controlled by genetic means and regulated by storage conditions. Kumar and Singhal observed that dominant maternal effect or some cytoplasmic genes govern the seed longevity.

### Seed Collections

Garden pea requires cool, relatively humid climatic conditions for optimum growth. It grows well in fertile, sandy loam soil with good drainage. It is a self-pollinated and long-day plant. An isolation distance of 20 m is maintained between varieties. Crops are provided with adequate fertilizer, irrigation, and plant protection measures for high yield. Seed quality and vigor are improved by application of nitrogen and phosphorus. Plants are checked especially during the flowering and pod development stages for various morphological characters, and off types are removed. Pods turn yellow on seed maturity. The ripe pods are collected and threshed, and the seeds are separated, cleaned, graded, dried, and stored.

### Seed Germination

Seed germination is hypogeal and rapid at alternate temperatures of 20/30°C. Fully mature seeds give better germination. Manohar and Sachan reported that seeds extracted 28 days after anthesis gave higher germination. Seed size determines the rate of seedling emergence. According to Edwards and Hartwig , small and medium-sized seeds give rapid seedling emergence and greater root development than large seeds, while Kant noted that large seeds are superior over smaller seeds for early field emergence.

Seed germination is delayed by the presence of hard seeds, as well as by slow imbibition of moisture in certain seeds. This can be overcome by exposing seeds to 40 MHZ radio frequency electric field or by soaking seeds in NAA (30 ppm) and GA (30 ppm).

### Storage Conditions

*Seed moisture.* Seed moisture plays an important role in maintenance of seed viability and vigor. Higher moisture is injurious to seed longevity. The optimum moisture for seed storage varies with temperature. Seed moisture of 6 to 7 percent is ideal for longer storage. Pavelkova and Curiova reported that rapid and excessive removal of moisture affects seed vigor, and also plant yield from such seeds. Drying of immature seeds, especially at higher temperatures, causes hollow heart, a physiological disorder of pea seeds. According to Agrawal and Gupta , hollow heart does not have any adverse effect on pea germination. Dressler noted that well-dried pea seeds stored at 10–1 to 10–2 mb maintained viability for 9 to 12 years.

*Temperature.* Seed viability and vigor are lost rapidly at higher temperatures. Kononkov, Kravchuk, and Vasyanova reported that seeds failed to germinate during the second month of storage when stored at 20 to 30°C and 82 percent RH, while at a lower temperature, seed deterioration is slow and seeds remain viable for a longer period. Seeds stored at 10°C and 50 percent RH remain viable for five years.

Likewise, when stored at subzero temperature (–20°C), seeds remained viable for 33 years.

*Storage fungi.* Seed health is affected by improper storage conditions. Fungi invade rapidly at higher seed moisture or at high relative humidity. Granett and Harman reported that seeds infected with *Aspergillus ruber* storage fungi show reduction in seed germination. This was controlled by the use of pentachloro (PCNB) nitrobenzene, thiram , and Ceresan and captan.

## Seed Deterioration

Changes in seed color and appearance; loss of weight, viability, and vigor; intolerance to unfavorable storage conditions; cellular disintegration; and inactivation of enzymes and DNA synthesis are associated with garden pea seed deterioration. Seed storage under extremely dry (1 percent RH at 10°C) or humid conditions (93 percent RH at 25°C) for six weeks reduced the germination percentage and increased solute leakage (sugar, potassium, electrolytes), and this was attributed to deterioration of the cell membrane. With an increase in storage period, the electrolyte leakage increases and total phospholipid content decreases. Seed viability is negatively correlated with the conductivity of leachates , and it is greater in low-vigor seeds.

The loss of viability is accompanied by a decline in nucleic acid levels, which includes a reduction in protein hydrolysis, protease activity, catalase activity, and peroxidase activity. The seed mesocotyl region, including the root and shoot meristem, seems to be the key tissue in pea to protect it from deterioration. This area must remain alive for germination. A few key cells in seeds have to remain functional if the seeds are to retain germinability.

These cells are involved with the transportation and assimilation of food. According to Harman, Khan, and Tao , aging in pea seeds is associated with slower oxygen uptake, a delay in protein synthesis, and reduced enzyme activity. Seed deterioration is slowed by osmotic priming of seeds with polyethylene glycol in acetone and storage at low moisture and low temperatures.

## Invigoration of Stored Seeds

Seed germination improves with presoaking of seeds in water or chemicals, especially in low-quality pea seeds, and this was attributed to the cellular repair mechanism.

# Chapter 53

# BROAD BEAN: *VICIA FABA* L.

## INTRODUCTION

Broad bean is also known by different names, such as favabean, horse bean, Windsor bean, field bean, and tick bean. It is cultivated in temperate, subtropical, and tropical regions for its immature and mature seeds. The seeds are cooked as a vegetable and also frozen or canned. The green crop is fed to cattle and also used as green manuring crop, which fixes the atmospheric nitrogen. Seeds are rich in proteins, carbohydrates, and oil and also contain high amounts of calcium and phosphorus.

### Origin and Distribution

Broad bean originated in the Mediterranean region. It is cultivated during the cooler season of the year. It is grown largely in Australia, Brazil, China, Ecuador, Egypt, England, Greece, India, Italy, Mexico, Myanmar, the Netherlands, and Sudan.

### Morphology

Broad bean belongs to family Leguminosae and has chromosome number 2n = 12, 14. It is an erect, dicotyledonous, annual herb that grows up to 1 m in height. Roots are well nodulated. Leaves are alternate and pinnate. The inflorescence is a short axillary raceme. Flowers are white, with purple patches, and fragrant. It is a self-fertilized plant. Fruit is a fleshy pod, 5 to 15 cm long, and contains one to six seeds. Seeds are large, brown, and flattened and vary in shape. One thousand seeds weigh about 800 to 2,000 g.

## SEED STORAGE

Seeds are used in the propagation of broad bean. They are also used in creating and conserving genetic variability. Seeds show orthodox storage behavior. They can withstand loss of moisture without affecting viability and vigor. These seeds also preserve high germination percentage during low-temperature storage.

### Seed Collections

Broad bean is a cool-season crop. It grows on high hills in the tropics. It prefers fertile, well-drained soil for cultivation. Seeds are directly sown in the field, and plants are supported during growth and development. It is largely a self-pollinated crop; however, cross-pollination occurs to a certain extent. Thus, an isolation distance of 500mis provided between varieties to safeguard against genetic contamination. Crops are provided with adequate nutrients, watered regularly, and protected against insect pests and pathogens (see Figure 1). Plants are checked during the flowering and fruiting stages, and off-type plants are removed. Pods become brown on maturity. Such pods are harvested when they become relatively dry, then threshed, and seeds are separated, cleaned, dried, and packed for storage.

### Seed Germination

*Figure 1. Broad Bean Vine with Many Mature Pods*

Seed germination is hypogeal and higher at a constant temperature of 25 to 30°C or at alternate temperatures of 20/30°C for 16/8 hours. Seeds take longer to germinate due to the presence of a hard seed coat. Germination can be improved by making a small incision on the seed coat or through scarification. Singh and Singh reported that seeds soaked in concentrated sulfuric acid for 30 min followed by washing with tap water enhanced the germination percentage. Longer acidification causes seedling injury. Moist stratification at 5°C also improves field emergence and promotes vegetative growth and flowering.

## Storage Conditions

Broad bean seeds are short-lived under ambient conditions. Seed deterioration is greater at high seed moisture and storage temperature. It is associated with loss of viability and reductions in root and shoot growth. At higher relative humidity, seeds gain moisture and lose viability rapidly, especially at higher temperatures. Seeds gave 94 percent germination when stored at room temperature and 76 percent RH, and zero at 30°C and 76 percent RH, after six months of storage, while viability was lost completely at 40°C and 76 percent RH after one month of storage. Seeds stored in open conditions maintain high viability, particularly at high temperatures. Jute and paper bags are ideal for storing high-moisture seeds. Seeds with moisture content greater than 19 percent showed 86 percent germination after six months of storage. Seed stored in liquid nitrogen (–196°C) preserve their initial viability for three years, while seed storage at –100 to –120°C results in low germination as compared to storage at –196°C. Fast thawing (10°C/second [sec]) is more effective than slow thawing (1°C/sec) in retrieval of high viability after ultralow-temperature storage of seeds.

# Chapter 54

# HYACINTH BEAN: *LABLAB NIGER* MEDIK.

## INTRODUCTION

Hyacinth bean is also called dolichos bean, Egyptian bean, Indian bean, or lablab bean. It is cultivated for its immature seeds and pods. Tender seeds and pods are cooked as a vegetable, while mature and dried seeds are used as pulses. Ripe and sprouted seeds are also cooked as a vegetable. Seeds are rich in protein and carbohydrates.

### Origin and Distribution

Hyacinth bean originated in India. It is widely cultivated in dry and lowrainfall regions. It is a popular vegetable crop in Brazil, China, Egypt, India, Indonesia, Malaysia, Papua New Guinea, the Philippines, and Sudan.

### Morphology

Hyacinth bean is a vigorous, bushy or twining, dicotyledonous, perennial herb. However, it is cultivated as an annual. It belongs to the family Leguminosae and has chromosome number 2n = 22, 24. Leaves are dark green, alternate, and trifoliate. Inflorescence is axillary, and flowers emerge in clusters of four or five. Flowers are hermaphrodite; white, red, or purple; and self-pollinated. Pod is long, oblong, and somewhat curved and contains three to five seeds. Seeds are round and of various colors, such as white, red, brown, black, or variously speckled, with a long white hilum. One thousand seeds weigh about 330 g.

## SEED STORAGE

Seeds are used for propagation and also for long-term conservation of genetic diversity. Seeds show orthodox storage behavior; they can withstand drying and maintain high viability under low-temperature storage conditions.

### Seed Collections

Hyacinth bean is a warm-season crop. It grows well in high-temperature and low-rainfall areas. It requires deep, fertile, well-drained soil and is susceptible to waterlogged conditions. Seeds propagate it. Seeds are sown directly in the field, and later, the plants are supported with poles. Crops are properly fertilized, irrigated, and protected against pests and diseases. Although hyacinth bean is a self-pollinated crop, insects bring about crosspollination to a certain extent. To prevent contamination, a seed plot is isolated from another variety by a 50 m distance. Plants are regularly checked, especially during the flowering and pod development stages, and off-type plants are removed. Pods become light yellow on maturity. It takes about 70 to 120 days from sowing to harvesting, depending upon the cultivar and growing conditions. Ripe pods are harvested, and seeds are separated, cleaned, dried, and suitably packed.

### Seed Germination

Seed germination is epigeal and completes in a week. There is no problem of dormancy. Seed germination is higher at a constant temperature of 25°C or at alternate temperatures of 20/30°C for 16/8 h. These seeds are sensitive to moisture stress during germination.

### Storage Conditions

Seeds are viable for two years under ambient conditions. Prevalence of high humidity and high temperature hastens the process of seed deterioration and reduces the storage life of seeds. Seeds coated with activated clay (1:100 by weight) retain high viability for 18 months under ambient conditions. Further, seed packaging in moisture-proof containers, such as aluminum cans or foil, maintains high viability and vigor at low (5°C) or subzero temperatures (–20°C).

Open storage of seeds, especially in high relative humidity areas, leads to the growth of fungi, which discolor seeds, increase temperature, and reduce viability. Seed treatment with captan (2 g/kg) preserves high viability and vigor and gives better field emergence after 40 months of storage under ambient conditions.

# Chapter 55

# CLUSTER BEAN: *CYAMOPSIS TETRAGONOLOBA* (L.) TAUB.

## INTRODUCTION

Cluster bean is a hardy pod vegetable that is commonly grown in the tropics and subtropics. It is cultivated for its edible pods. Tender pods are cooked as a vegetable. The whole plant is fed to cattle and also used for green manuring purposes. Seeds are rich in protein, carbohydrates, and mannogalactans; the latter have several industrial usages, such as in the textile, paper, and cosmetic industries.

### Origin and Distribution

Cluster bean originated in India. Wide genetic diversity for various plant characters, fruit characters, and chemical constituents is found in different parts of India. It is cultivated in Brazil, India, Indonesia, Malaysia, Pakistan, the Philippines, and Sri Lanka.

### Morphology

Cluster bean belongs to family Leguminosae and has chromosome number 2n = 14. It is a dicotyledonous, herbaceous, annual, bushy plant. Leaves are medium size, dark green, and trifoliate. Flowers are small, white or pink, and borne on an axillary raceme. Pods are linear, long, compressed, green, erect, beaked, and borne in clusters. Fruit contains 5 to 12 seeds that are 5 mm long; white, gray, or black; and quadrangular in shape. One thousand seeds weigh roughly 25 to 51 g.

## SEED STORAGE

Cluster bean seeds are short-lived under ambient conditions. Seeds are used for commercial cultivation and also as a valuable tool in long-term conservation of genetic diversity. Cluster bean seeds exhibit orthodox storage behavior; they can withstand desiccation and chilling temperatures during storage.

### Seed Collections

Cluster bean is a warm-season crop. It grows well in dry conditions and withstands drought. It requires deep, sandy loam soil for successful cultivation. It is grown from the seeds, which are sown directly on ridges in the field. It is a self-pollinated crop, and about 50 m of isolation distance is maintained during seed production. Normal cultural practices, such as spacing, fertilization, irrigation, and plant protection measures, are followed. Short-day conditions are favorable for flowering. The crop is inspected during the flowering and pod development stages, and off-type plants are removed. Seeds mature about 150 days after planting. Pods became light yellow and somewhat hard on maturity. The ripe dry pods are collected and threshed carefully, without injuring the seeds. Seeds are dried, cleaned, and suitably packed for storage.

### Seed Germination

Seed germination is rapid and higher at a constant temperature of 30°C and at alternate temperatures of 20/30°C for 16/8 hours. Germination is epigeal and completes within a week. Kalavathy and Vanangamudi reported that large seeds gave higher germination and seedling vigor than the smaller seeds. During germination, total water-soluble carbohydrates and dry matter contents decrease. Seed germination is also affected by the use of certain insecticides, such as monocrotophos, on dry seeds.

### Storage Conditions

Cluster bean seeds survive for a short period at ambient temperatures. Seed quality is affected by improper storage conditions, such as presence of high temperature and relative humidity. Seeds stored in kraft paper bags maintain high viability for 12 months, while those in sealed storage in glass, polyethylene bags, or laminated aluminum foil pouches retain viability relatively longer under ambient conditions . Seeds showed 93 percent germination after 15 years of storage at 5 and –20°C. There was a slight reduction in the viability of seeds stored in polyethylene bags, but no morphological variations were recorded in seedlings raised from these seeds. On seed aging, the germination percentage decreases, and the leakage of electrolytes, soluble sugars, and free amino acids increases on seed imbibitions.

# Chapter 56

# WINGED BEAN: *PSOPHOCARPUS TETRAGONOLOBUS* DC.

## INTRODUCTION

Winged bean is widely cultivated in humid tropical and subtropical regions. It is also known as four-angled bean, goa bean, Manila bean, and winged pea. Almost all plant parts are used for human consumption. Tender shoots, tuberous roots, young leaves, flowers, immature pods, and mature seeds are cooked as vegetables. Plants are also fed to cattle and used as a green manure. Winged bean is rich in protein, fats, carbohydrates, calcium, phosphorus, and iron.

### Origin and Distribution

Winged bean appears to have originated in Southeast Asia, probably in Papua New Guinea. It is cultivated in Ghana, India, Indonesia, Malaysia, Myanmar, Nigeria, Papua New Guinea, the Philippines, Sri Lanka, Tanzania, Thailand, and Vietnam.

### Morphology

Winged bean belongs to family Leguminosae and has chromosome number 2n = 26. It is a perennial herb but is cultivated as an annual. The tuberous root is well nodulated, fixes atmospheric nitrogen, and enriches the soil. Leaves are dark green and trifoliate. The inflorescence is an axillary raceme that bears two to ten flowers. Flowers are white, light blue, or purple. Fruit is a green pod, having four sides with wings, which vary in size, and it contains 8 to 17 seeds. Seeds are globular, smooth, and shiny and vary in color, being either white, yellow, brown, or black. One thousand seeds weigh about 500 g.

## SEED STORAGE

In winged bean, seeds are used in propagation, in crop improvement through hybridization, and in long-term genetic conservation. Seeds are short-lived and remain viable for one or two years at room temperatures. Seeds show orthodox storage behavior; their viability is not affected by drying and chilling temperatures during storage.

### Seed Collections

Winged bean grows well in hot humid conditions. It requires deep, fertile, well-drained soil. Crops are sensitive to waterlogging. Seeds are directly sown in the field, and the plants are supported by poles. Normal cultural practices, such as fertilization, irrigation, and plant protection measures, are followed. Flowering commences about 50 days after sowing. Winged bean is a self-pollinated crop; however, cross-pollination by bees does occur to a certain extent. An isolation distance of 50 m is maintained between cultivars, and off-type plants are removed. Seeds mature 60 days after anthesis. Pods become brown on maturity. At this stage, pods are collected and dried, and the seeds are extracted.

*Seed maturity.* Seed maturity determines viability and longevity. Immature seeds exhibit low vigor and viability. Abenoja noted that seeds harvested in the early days of pod development (5 to 20 days after flowering) showed lowest viability and vigor, compared to late-harvested seeds (25 to 35 days after flowering), which also gave higher germination and better seedling growth. Seeds attain maximum dry weight on the thirtieth day after anthesis.

### Seed Germination

Seed germination is hypogeal, and seedlings take longer to emerge from the soil. Hew and Lee reported that seed germination and seedling growth are better at a constant temperature 25°C and in dark conditions. Seedling emergence is uneven in the field, and this is attributed to the impermeability of the seed coat. Uneven emergence was overcome through pregermination softening of the seed coat. This includes testa abrasion by sandpaper and acid scarification. Seed soaking in dilute sulfuric acid (8:1 by volume in water) for 5 min at 62°C followed by rinsing in running water for 5 min gave 96 to 100 percent germination, without any injury to emerging seedlings.

## Storage Conditions

Winged bean seeds lose their viability rapidly at higher temperature and higher relative humidity. Johnson and Madhusudanan noted that none of the seeds germinated after 24 months of storage under ambient conditions. Seed germination decreased to 50 percent when fresh seeds (20 to 21 percent mc) were stored for two weeks under ambient conditions. Further, seeds lost their viability completely after two months of storage. High-moisture seeds (12.5 to 18.5 percent) stored at 8°C retained seed color, appearance, and high viability for 24 weeks.

# Chapter 57

# COWPEA: *VIGNA UNGUICULATA* (L.) WALP.

## INTRODUCTION

Cowpea is cultivated in the tropics for its edible tender long pods and dry seeds. It is also called Bombay cowpea or catjang. Immature pods and fresh seeds are cooked as a vegetable or canned. Green foliage is fed to the cattle. Cowpea is rich in protein, carbohydrates, oils, calcium, phosphorus, and vitamin A.

### Origin and Distribution

Cowpea originated in Asia. Wide genetic diversity, including wild relatives, exists in India and China. Major cowpea-producing countries are China, Ghana, India, Indonesia, Malaysia, Nigeria, the Philippines, Sri Lanka, and the United States.

### Morphology

Cowpea belongs to family Leguminosae and has chromosome number 2n = 22. It is a dicotyledonous, annual, bushy or climbing herb. Roots con tain a large number of nodules. Stems are erect or semierect with trifoliate leaves. The inflorescence is axillary, with two to four flowers in a cluster.

Flowers are yellow or violet, perfect, and borne on a long peduncle. Pods are long, thin, white, or green and contain 10 to 30 seeds. Seeds vary in size and shape, from globular to cylindrical. One thousand seeds weigh about 220 g.

## SEED STORAGE

Cowpea is cultivated by seeds. Seeds are used in breeding programs for crop improvement, in long-term genetic conservation, and as a valuable source of proteins. Cowpea seeds show orthodox storage behavior. These seeds tolerate drying, and initial viability can be maintained at low temperatures. Cowpea can survive for only a short period under ambient conditions due to heavy infestation by insects.

### Seed Collections

Cowpea is a warm-season crop. It prefers a hot climate for optimum cultivation. It is sensitive to cold and tolerates heat and dry conditions. It grows well in deep, fertile, sandy loam or clay soils with good drainage. Crops are raised by seeds and sown directly in the field. It is self-fertilized plant; however, insects bring about cross-pollination to a certain extent. To safeguard against genetic contamination, an isolation distance of 50 m is maintained between varieties. Crops are provided adequate nutrients, irrigation, and plant protection measures for ideal growth. Plants are checked during the vegetative and reproductive phase, especially during the flowering and fruiting stages, and off-type plants are removed. Pods take about 90 to 150 days to mature, and they become straw colored on maturity. Dry pods must be harvested in time; otherwise, delay leads to shattering of seeds. Pods are threshed, and seeds are separated, cleaned, dried, and packed for storage.

*Fruit maturity.* Fruit maturity almost coincides with that of seed maturity. Fruit takes about 90 to 150 days to mature, depending upon the cultivar and growing conditions. Cowpea seeds develop rapidly during the earlier period of postfertilization. There is continuous accumulation of dry matter in the seeds, and this ceases on the attainment of mass maturity. Fully matured seeds have high viability and vigor during storage. At this stage, pods change color and become somewhat brittle. The husk development is completed earlier than that of other pod parts. Manohar and Mathur reported that seeds attain germination capability in six days after the opening of flowers. However, the germination percentage and seedling vigor are greater for seeds harvested at 14 to 20 days after pollination. Pods could be harvested at the color change stage, with low viability and vigor, to avoid unforeseen unfavorable circumstances, such as bird damage, excess rainfall, and outbreak of other biotic and abiotic stresses. Early harvested seeds (12 days after anthesis) retain viability for six months at room temperature, as compared to more than a year in seeds harvested at 16 to 20 days after flowering.

### Seed Germination

Seed germination is epigeal and completes in a week. However, there will be uneven emergence of seedlings due to the presence of hard seeds, which increases with a decrease in water potential. Soil temperature of 25 to 30°C is suitable for better germination. In the laboratory, seed germination hastens at alternate temperatures of 20/30°C for 16/8 h, respectively. *Seed size.* In cowpea, seed size varies among and within cultivars, based on seed position in the fruit and growing conditions. Shashidhar, Vyakarnahal, and Swamy reported that large seeds were more vigorous and gave higher percentage of germination. This was attributed to better food reserves in seeds. Large seeds remain viable longer than smaller seeds.

## Storage Conditions

Cowpea seeds are short-lived under ambient conditions. Rodrigo reported that seeds remain viable for nine to ten months in open storage. Cowpea seeds are easily attacked by insects; thus, a combination of a good physical storage environment and pest-free healthy seeds is required to maintain high seed viability during storage. Durvasula noted that storage conditions of 20°C and 45 percent RH are ideal for short-term storage of cowpea seeds. Solar heaters are reported to be beneficial in reducing insect activity and in creating dry conditions during ambient storage.

*Seed moisture.* High seed moisture is injurious to seed longevity. Seeds with 11.7 percent moisture maintain their viability up to 300 days under ambient conditions.

Further, seed drying promotes hard-seededness, and moisture below 6 percent affects germination. Vanangamudi reported that low-moisture seeds (8 percent) preserved their viability for 40 months under ambient conditions.

*Seed packaging.* Seeds are packed to facilitate their handling during storage, and also to create a small storage environment wherein the desired level of moisture or relative humidity can be maintained for a fairly long period. Different types of containers are used according to storage period and storage conditions. Figueiredo, Frazao, Oliveira, and Correa reported that retention of seed viability was less in paper bags than in cloth and polyethylene bags. Low-moisture seeds are stored in moisture-proof containers such as aluminum foil or cans for better and longer retention of seed quality.

*Temperature.* High-temperature storage promotes seed deterioration, thereby reducing seed vigor. Seeds lose their viability rapidly at room temperatures and thus store better under refrigeration. Insect activity is high in high-moisture seeds at higher temperatures, thereby affecting seed quality.

*Storage pests and fungi.* Cowpea seeds are severely damaged by pulse beetle (*Callosobruchus chinensis* L.) during storage. Insect activity is rapid in high-moisture seeds, at high temperatures, and in the later period of storage. Application of ethylene dibromide as a fumigant, malathion (5 percent) , or neem oil (1:100 parts) effectively controls pulse beetle infestation. Use of a solar heater during short-term storage of cowpea seeds reduces insect infestation by providing heat sufficient to kill all developing insects inside the seeds.

The occurrence of heavy rains during harvest increases seed mycoflora and decreases germination percentage. *Aspergillus* spp., which are present externally, and *Fusarium moniliforme* and *macrophomina phaseolina,* which are present internally, affect seed germination. Fungicides such as Ceresan, thiram, carbendazim, and captan are effective in controlling the pathogens and help in maintaining high viability and vigor during storage. Mancozeb and seed coating with red earth (10 g/kg) or activated clay eliminated the storage fungi.

# LEAFY VEGETABLES

# Chapter 58

# FENUGREEK: *TRIGONELLA FOENUM-GRAECUM* L.

## INTRODUCTION

Fenugreek is cultivated for food, fodder, and green manure in the tropics and subtropics. Its green leaves are commonly cooked as a vegetable. Seeds contain proteins, carbohydrates, and oils. Seeds are used as spices in culinary purposes and have medicinal properties. Fenugreek is a rich source of vitamins A and C.

### Origin and Distribution

Fenugreek is native to Asia and southern Europe. It is cultivated in Egypt, Ethiopia, India, Indonesia, Italy, Malaysia, and Russia.

### Morphology

Fenugreek belongs to family Leguminosae and has chromosome number 2n = 16. It is a quick-growing dicotyledonous plant having light to dark green trifoliate leaves. Flowers are small and white. Pods are long, slender, and straw colored and contain 8 to 15 seeds. Seeds are smooth and yellowish brown. One thousand seeds weigh about 10 to 12 g.

## SEED STORAGE

Seeds are commonly used in fenugreek cultivation. They are also used in crop improvement and conservation of genetic diversity. Seeds exhibit orthodox storage behavior; they withstand moisture loss and can be stored at chilling temperatures.

### Seed Collections

Fenugreek is a cool-season crop. It grows well in moderate- or low-rainfall areas. It prefers fertile, sandy loam soil. It is propagated by seeds, which are sown directly in the field. Plants are provided with adequate nutrients, irrigation, and plant protection measures. Plants spaced 20 cm apart give higher seed yield than those spaced 40 cm apart. It is a self-pollinated crop, and an isolation distance of 50 m is maintained between cultivars. Plants are checked during the flowering and fruiting stages, and off-type plants are removed. Matured pods are collected, dried, and threshed, and the seeds are separated, cleaned, dried, and suitably packed for storage.

### Seed Germination

Seed germination is epigeal and higher at a constant temperature of 20°C (Singh, Singh, and Singh, 1990). Fresh seeds exhibit dormancy under natural conditions, even after prolonged storage, due to the presence of a hard, thick seed coat. Seed germination is higher at alternate temperatures of 20/30°C for 16/8 h. Sinha, Jha, and Varshney reported that seed scarification for 2 or 4 min was effective in reducing the number of hard seeds. Seed germination is affected by saline conditions and promoted by gibberellic acid (10 mg·liter –1) application.

### Storage Conditions

Higher seed moisture and higher storage temperature promote seed deterioration. Well-dried seeds retain their viability for a longer period under cooler conditions. Healthy, bold seeds are selected and dried to 5 to 7 percent moisture for storage. These are suitably packed in polyethylene bags or in aluminum foil pouches and stored at low temperatures for longer storage life.

*Storage fungi.* Storage fungi grow fast when seed moisture is high, thereby affecting seed quality. Fungi such as *Alternaria, Aspergillus, Cladiosporium, Fusarium, Penicillium, Phorma,* and *Rhizoctonia* associate with fenugreek seeds. Seed dressing with captan, carbendazim, and thiram controls the fungi. Boric acid is also effective in eradicating them.

# Chapter 59

# SPINACH: *SPINACIA OLERACEA* L.

## INTRODUCTION

Spinach is an important winter leafy vegetable cultivated for its smooth or wrinkled edible leaves. It is cooked as a vegetable and used raw in salad. It contains protein, carbohydrates, calcium, iron, and vitamins A, B, and C.

### Origin and Distribution

Spinach originated in Central Asia. It is predominantly grown in temperate regions; however, it is also cultivated at higher altitudes in the tropics and subtropics. It is a popular leafy vegetable in Canada, Denmark, Egypt, France, Germany, the Netherlands, Poland, Russia, Spain, the United Kingdom, and the United States.

### Morphology

Spinach belongs to family Chenopodiaceae and has chromosome number 2n = 12. It is a dicotyledonous, dioecious, erect, herbaceous biennial plant with edible leaves. The inflorescence is a terminal panicle; male flowers are small and open earlier into pistillate flowers. Female flowers are axillary and greenish and number 6 to 12 per cluster. Fruit is single seeded. Seeds are round and smooth or prickly. One thousand seeds weigh about 10 g.

## SEED STORAGE

Seeds are widely used in crop production. They are also used for evolving new varieties and in the conservation of genetic diversity. Seeds show orthodox storage behavior and are fairly long-lived under ambient conditions.

## Seed Collections

Spinach is a cool-season crop that grows well in low-temperature areas. High temperature induces premature bolting. Young seedlings can withstand frost. Spinach is a long-day plant. Flowering is profuse under long-day and chilling conditions. The plant bears staminate, pistillate, or hermaphrodite flowers separately. It grows well in fertile, well-drained, sandy loam soil.

Crops are fertilized, irrigated, and sprayed regularly against insect pests and diseases. Stems elongate and form flowers in clusters. Spinach is cross-pollinated by the wind; therefore, an isolation distance of 1,600 m is provided between varieties. Roguing is done during the flowering and fruiting stages, and off-type plants are removed. On maturity, plants are cut and threshed, and the seeds are separated, cleaned, dried, and packed for storage.

## Seed Germination

Spinach seeds show a considerable amount of dormancy. Prevalence of low temperature also induces dormancy. Such dormant seeds require afterripening as well as higher temperatures for germination. Khalilov reported that 21 days of after-ripening was beneficial for eliminating dormancy. Seeds germinate at 5 to 27°C, and the optimum temperature is between 20 and 25°C. Spinach seeds can also tolerate a high level of soil salinity. Seed germination is hastened by exposing seeds to a 40 MHZ radio frequency electric field.

## Storage Conditions

Seed storage under improper conditions affects seed quality. Seed viability is lost rapidly in high-moisture seeds and also at high storage temperature and relative humidity. Nath (1976) reported that spinach seeds remain viable for three to four years under proper storage conditions. Seeds having 5 to 7 percent moisture that are suitably packed in polyethylene bags and in aluminum foil pouches or cans and stored at lower temperatures maintain high viability and vigor during storage.

# Chapter 60

# AMARANTH (*AMARANTHUS* SPP.)

## INTRODUCTION

Amaranth is a popular leafy vegetable of the tropics. Different species thatare cultivated predominantly as vegetables are *Amaranthus cruentus* L. in Africa, *Amaranthus tricolor* L. in Asia, and *Amaranthus dubius* Mart. Ex Thellung in the Caribbean. In the Americas, the varieties of *Amaranthus hypochondriacus, A. cruentus,* and *A. caudatus* are largely cultivated for grain purposes. Amaranth is cultivated for its edible leaves, which are cooked as a vegetable. Also, a few cultivars are grown exclusively for grain purposes, and these grains are used for food purposes. Amaranth is rich in proteins, magnesium, phosphorus, sodium, calcium, iron, sulfur, potassium, and vitamins A and C. It also contains a high amount of oxalic acid.

### Origin and Distribution

The origin of amaranth is controversial. Genetic diversity exists in both the New World and the Old World, predominantly in South America and South Asia. Amaranth as a leafy vegetable is mainly grown in Brazil, China, Ethiopia, India, Indonesia, Kenya, Malaysia, Mexico, Myanmar, the Philippines, Taiwan, Uganda, and the United States.

### Morphology

Amaranth belongs to family Amaranthaceae and has chromosome number 2n = 32. It is a dicotyledonous, monoecious, short-lived annual plant. Stems are straight, grow up to 1 m in height, and end with terminal inflorescences. Leaves are opposite and green or purple in color. The inflorescence is an axillary or terminal racemose that is green, orange, or purple and contains numerous staminate and pistillate flowers. Flowers have four to five petals, one to five stamens, and two or three carpels with superior ovaries. Fruit is an indehiscent utricle surrounded by a persistent perianth. Seeds are small, black, brown, or white and shiny. One thousand seeds weigh about 0.3 g.

## SEED STORAGE

Seeds are widely used for cultivation, for crop improvement, and in the long-term conservation of genetic diversity. Amaranth seeds show orthodox storage behavior. They remain viable for a short period under ambient conditions. Seed storability increases with drying and storage at low temperatures.

### Seed Collections

Amaranth is a warm-season crop and needs warmer conditions for optimum growth. It grows well in hot, sunny places. Seeds are directly sown in the field. Light soil with good drainage is ideal for cultivation. Often the terminal growing point is cut to encourage side shoots. The crop is wind pollinated, but selfing is more common. An isolation distance of 500 m is maintained. Roguing is done to remove off-type plants. Rajan and Joseph reported that seed quality was unaffected by cutting the leaves for vegetable purposes. Spikes are collected when glumes turn brown and seeds become black. Subsequently, spikes are dried and threshed, and the seeds are separated. Seeds are dried and cleaned, and the good seeds are selected for storage.

### Seed Germination

Amaranth seeds show a certain amount of dormancy. Seed storage for short duration eliminates dormancy. Seed germination rate decreases due to low soil temperature. Temperature of 30°C is optimum for better seedling emergence. The process of seed germination is enhanced by exposure of seeds to alternate temperatures (20/30°C for 16/8 h). Aufhammer and colleagues noted that light or short illumination inhibits the germination process.

### Storage Conditions

Seed viability is affected by improper storage conditions. Seed viability is reduced after one year of ambient storage. Seed viability is preserved by packing seeds in polyethylene bags and in aluminum foil pouches and storing them at 5 and –20°C for ten years. Storage of seeds in nitrogen maintains higher germination and seedling vigor for seven years.

# Chapter 61

# CORIANDER: *CORIANDRUM SATIVUM* L.

## INTRODUCTION

Coriander is cultivated as spinach and mainly used in flavoring food dishes. Dry seeds are powdered and used as a spice in food preparation. It is rich in proteins, carbohydrates, calcium, iron, and vitamin A. Fruit contains a volatile aromatic oil that is used in medicines and has carminative properties.

### Origin and Distribution

Coriander originated in the Mediterranean and Middle East regions. It is cultivated in Brazil, China, England, France, Germany, India, Italy, Malaysia, Morocco, Thailand, and the United States.

### Morphology

Coriander belongs to family Umbelliferae and has chromosome number 2n = 22. It is an annual, dicotyledonous herb that grows up to 80 cm in height. The stem is hollow, with broad leaves at the base and narrow leaves at the top. The inflorescence is a terminal umbel with numerous small white, pink, or lavender flowers. The fruit is globular, yellow, or brown; ribbed; and about 4 mm in diameter. It is a two-seeded schizocarp. Seeds are semiglobular and contain essential oil (0.1 to 1 percent), mainly coriandrol (C10 H18 O). One thousand seeds weigh roughly 7 to 13 g.

## SEED STORAGE

Coriander is propagated by seeds. It is also used in breeding programs and in long-term conservation of germplasm. Seeds show orthodox storage behavior. Seed viability is unaffected by drying and low-temperature storage.

### Seed Collections

Coriander needs a cooler climate for optimum growth, especially at higher altitudes. It grows well in deep, fertile, well-drained soils. Seeds are sown directly in the field. Normal cultural practices are followed for high crop production. It is a cross-pollinated crop, and about 1,000mof isolation distance is maintained to prevent genetic contamination. Off-type plants are removed. Fruit requires about 90 to 120 days from sowing to maturity. Dry, mature fruits are collected, separated, and cleaned for storage.

### Seed Germination

Fresh seeds do not germinate readily, revealing dormancy. Seed germination is better at 15°C, and it takes 10 to 21 days to complete. Seed germination decreases above 30°C and lower than 10°C. Seeds selected from the second-order umbel gave higher germination and higher seedling vigor than those from the first- or third-order umbels. Germination is improved by subjecting seeds to alternate temperatures of 27/22°C for 8/16 h, respectively.

### Storage Conditions

Seed viability and vigor decrease with an increase in seed moisture and storage temperatures. Low-moisture seeds are packed in polyethylene bags or in laminated aluminum foil pouches and are stored at low temperature. This maintains high viability for a longer period.

*Storage pests and fungi.* Higher seed moisture is congenial for insect and pathogen activity. Verma reported that fumigation with methyl bromide (64 ml·liter −1) or hydrogen cyanide (HCN) (48 mg·liter −1) gas for 2 h at reduced pressure kills the insects inside and does not affect seed quality.

Fungal damage is controlled by application of fungicides, such as brassicol, zineb, captan, and thiram, at 0.3 to 0.4 percent.

# SALAD VEGETABLES

# Chapter 62

# LETTUCE: *LACTUCA SATIVA* L.

## INTRODUCTION

Lettuce is a popular salad crop in temperate regions. It is a cool-season vegetable crop exclusively cultivated for its edible leaves. It is also grown to a lesser extent and on a smaller scale in the subtropics, where it is cooked as a vegetable. It is rich in minerals such as calcium, phosphorus, and iron and vitamins A and C.

### Origin and Distribution

Lettuce originated in the Middle East region composed of Iran and Turkey, where the wild species are found. It might have evolved from *Lactuca serriola* L. Lettuce is cultivated in Australia, Canada, China, France, Italy, Malaysia, the Netherlands, the Philippines, South Africa, Spain, the United Kingdom, and the United States.

## SEED STORAGE

Lettuce seeds show orthodox storage behavior. Seed viability is unaffected by loss of moisture and by exposing seeds to chilling temperatures. Seeds are commonly used in commercial propagation and for long-term genetic conservation.

### Morphology

Lettuce belongs to family Compositae and has chromosome number 2n = 18. It is cultivated as an annual for vegetable purposes and as a biennial for seed purposes. Lettuce is a dicotyledonous herb with a small stem that contains latex. Leaves are alternate, sessile, and spirally arranged in a rosette. Outer leaves are dark green and rough, whereas the inner ones are light green and succulent. Heads vary in size and shape and are classified as cos, butter head, or crisp and leafy type. Inflorescence is a capitulum that contains yellow perfect flowers arranged in a cluster. Fruit is a singleseeded achene, black, brown, or white in color. One thousand seeds weigh about 0.6 to 1.0 g.

## Seed Collections

Lettuce is a cool-season crop. It grows well under temperate conditions and requires humus and well-drained soil. Dry conditions or water shortage induces premature bolting. It is propagated by seeds. Crops are provided with optimum dosages of nutrients, regular irrigation, and plant protection measures for the production of high-quality seeds. Lettuce is mostly a selfpollinated crop; however, cross-pollination occurs to a small extent. An isolation distance of 50 m is maintained between varieties to safeguard against genetic contamination. To facilitate flowering, a slight cut is made on the head or the upper leaves are removed. Off-type plants are removed during the flowering and fruiting stages. Fruits take two weeks to mature after anthesis and are harvested when the capitulum bursts and shows the pappus. Seeds shatter if harvesting is delayed. Plants are cut and placed in small piles, dried, and threshed. Seeds are separated, cleaned, and packed for storage.

## Seed Germination

Lettuce seeds germinate readily and do not show any dormancy. However, seeds exposed to conditions of extreme temperature or light show secondary dormancy. The optimum temperature for seed germination is 25°C, and germination takes four or five days to complete. Seeds do not germinate when exposed to higher temperatures. Damania reported that a temperature regime greater than 25°C inhibits the germination process, due to the impermeability of the seed coat to oxygen. Continuous exposure of seeds to a dark period also induces secondary dormancy (skotodormancy).

*Light.* Light plays a vital role in lettuce seed germination. It promotes germination activity under a set of conditions. Sarma and Chakraborthy reported that germination rate is higher in the presence of light than in darkness. The exposure of moist seeds to fluorescent light for 24 h alone and/or followed by 24 h of darkness accelerates the germination process. The uneven emergence of seedlings caused by higher temperature is overcome by exposing moist seeds to red light for 3 h , and this eliminates the secondary seed dormancy. Globerson reported that while red light improves germination, far-red light inhibits it.

*Chemicals.* Seed germination improves by imposing certain chemical treatments, such as osmoconditioning with PEG at 10 bar, acid immersion for an hour followed by washing , and application of ethephon , gibberellic acid (50 mg·liter–1) or kinetin, thiourea (1,000 mg·liter–1) , and vitamin K (50 ppm). Germination is affected by volatile toxic substances released within storage containers.

Hsiao, Vidaver, and Quick (1984) reported that immersion of skotodormant seeds in acid followed by either gibberellic acid application or exposure to red light improved germination. Acid treatment weakens the membrane barrier of the endosperm cells and results in better penetration by gibberellic acid, thereby inducing higher germination.

*Storage of pregerminated seeds.* Pregerminated seeds are viable for a very short period. However, they can be stored up to five days at 5°C in aluminum foil pouches. Further storage reduces vigor and viability due to degradation of the cellular structure and inactivation of enzymes.

## Storage Conditions

Seed moisture or relative humidity, storage temperature, and oxygen play a major role in deterioration of lettuce seeds. Proper and careful regulation of these parameters is beneficial in retention of high viability on long storage. Lettuce seeds show good viability even after 15 years of storage.

*Seed moisture.* Seed moisture is reduced with the development of seeds. It remains constant at the physiological maturity of seeds. Further, moisture is reduced by drying seeds to a safe limit for better storage. Lettuce seeds deteriorate rapidly under high humidity and temperatures.

Seeds gain or lose their moisture from or to the atmosphere based on changes in vapor pressure. Seeds stored at fluctuating humidities at 5°C or 0 percent RH at room temperature maintain viability for four years. At highest and lowest humidity, dormancy developed in seeds and was apparently associated with seed deterioration, leading to death. The best seed viability is maintained at 46 to 58 percent RH. Kraak and Vos ( observed more abnormal seedlings when seed moisture exceeded 12 percent, and seeds were killed within a day when moisture exceeded 47 percent.

Excessive drying also affects seed quality. Crop yield decreased in seeds stored at 4.1 percent mc compared to those at 6.4 percent mc. Dressler  preserved well-dried seeds at 10–1 to 10–2 mb for 9 to 12 years without loss of viability.

*Seed packaging.* Packaging creates a barrier between seeds and the external environment. It protects the seeds from high relative humidity and prevents the entry of pathogens. The selection of packaging depends on the duration of storage. Normally, low-moisture seeds are kept in sealed storage.

Seeds stored in thinner-gauge bags lose their viability rapidly and exhibit more abnormalities than those stored in thicker-gauge bags. Seeds stored in polyethylene bags at –20°C lost moisture from 6.6 to 5.6 percent in 11 years of storage, and there was no reduction of seed viability during the period.

*Desiccants.* Chemical desiccants are effective in lowering seed moisture, thereby extending storage life. High seed quality was maintained for 86 months when lettuce seeds were stored with desiccants under fluctuating temperatures (5 to 30°. This method is quite beneficial for short-term storage, easy to manage, and inexpensive.

*Temperature.* Seed longevity decreases with increasing temperature. Seeds remain viable for four years at 5 or 10. At 20°C, viability is lost more rapidly in dry seeds (5.5 to 6.8 percent mc) than in ultradry seeds (2.0 to 3.7 percent mc). Ultradry seeds maintained viability for five years at –20°C storage.

*Air.* High oxygen content affects seed quality during storage. Harrison and Mcleish reported that lettuce seeds retain high viability in sealed containers when stored with carbon dioxide, which maintains higher germination, reduces chromosomal aberrations, and delays aging of seeds. Rao and Roberts opined that low-moisture seed survival was better in nitrogen than in air, vice versa for high-moisture seeds.

## Invigoration of Stored Seeds

Seed quality declines during ambient storage. Midstorage hydrationdehydration of medium-vigor lettuce seeds effectively reduces physiological deterioration and gives better field emergence and higher yields, whereas soaking-drying is not effective in higher-vigor seeds.

To improve germination, seeds are to be soaked in water for 2 h followed by storage at 4.4°C for four to six days. Seed soaking in thiourea (0.5 percent) also improves the germination percentage. The hydration-dehydration treatment curtails the process of deterioration and minimizes chromosomal aberrations. Introduction of chemicals such as ethephon (30 ppm) , potassium iodide (10–4 M), p-hydroxybenzoic acid (10–4 M), or tannic acid (10–5 M) into the seeds by dry permeation technique, using acetone as a solvent on the old seeds, re duces seed deterioration and promotes germination by counteracting the effect of free-radical damage to the stored seeds.

*Moist storage.* Higher seed moisture causes a rapid decline in longevity and also produces chromosomal aberrations. Seeds stored under an imbibed state do not germinate, which helps in the retrieval of high seed quality. Powell, Leung, and Bewley reported that imbibed seeds held at 25°C in darkness retained viability up to ten months. Seeds behave differently at a higher level of moisture (15 percent), and oxygen is essential for survival.

At this stage, the relative effect of temperature on decreasing longevity is slightly diminished. Seeds at 20 to 30 percent moisture do not suffer further decreases in longevity. It appears that hydration activates the repair mechanism. Seedlings grown from dry seeds stored under ambient conditions showed an increase in morphological abnormalities with an increase in storage period, whereas seedlings from imbibed stored seeds appeared normal. The repair mechanism operates only under the imbibed state, under the conditions that prevent germination during storage.

# Chapter 63

# CELERY: *APIUM GRAVEOLENS* L.

## INTRODUCTION

Celery is an important salad crop in Europe and North America that is grown for its edible leaf stalks and swollen roots. It is consumed raw or cooked as a vegetable. In temperate regions, the petioles are eaten as a fresh vegetable after blanching or flavoring. Dry leaves are used as a flavoring agent and in medicine. Seeds contain 2 to 3 percent volatile oil. Celery is rich in minerals such as calcium, phosphorus, and iron and in vitamin A.

### Origin and Distribution

Celery originated in Asia and Europe. It is largely cultivated in Europe and North America. Some of the major celery-growing countries are Algeria, Canada, China, Egypt, France, Indonesia, Malaysia, New Zealand, the Netherlands, the Philippines, Sweden, the United Kingdom, the United States, and the West Indies.

### Morphology

Celery belongs to family Umbelliferae and has chromosome number 2n = 22. It is a dicotyledonous biennial crop. It completes a life cycle in one year under temperate conditions. Celery roots are thick and fleshy; leaves are pinnately compound with long petioles up to 40 cm. Foliage is light or dark green and possesses a strong flavor. The inflorescence consists of small white flowers in an umbel. Flowers are perfect and self-fertile. Fruit is a rigid schizocarp that contains one seed. Seeds are small, brown, and ribbed. One thousand seeds weigh about 0.5 g.

## SEED STORAGE

Celery is propagated by seeds. Seeds are important in genetic conservation and crop improvement. Seeds are moderate storers under ambient conditions. They store well under dry, cool climatic conditions.

## Seed Collections

Celery is a cool-season crop that grows well in low-temperature regions. It requires deep, fertile, well-drained, humus soil. It is raised by seeds in the nursery and transplanted in the field. Seeds are soaked in water before sowing for early and higher germination. Plants undergo vernalization for the induction of flowering. Celery is a self-fertile plant, but insects bring about cross-pollination to a limited extent. Thus, an isolation distance of 500 m is provided between varieties. Adequate horticultural practices, such as irrigation, nutrition, and crop protection measures, are followed. Crops are inspected during the vegetative and reproductive phases, and off-types plants are removed. Fruits become brown on maturity, and at this stage, crops are harvested. Delay in harvesting causes shattering of seeds. Later, the plants are threshed, and the seeds are separated, dried, and suitably packed for storage.

## Seed Germination

Seed germination is epigeal and higher at alternate temperatures of 15/22°C. In celery, the seed embryo is comparatively underdeveloped, and embryo size varies in different cultivars. According to Flemion and Uhlmann , about 20 percent of seeds are empty. Further, immature embryos are dormant. All these factors contribute to delayed and uneven emergence of seedlings. Seed dormancy is also dependent on the umbel order. Thomas, Gray, and Biddington  reported that seeds from primary and secondary umbels were large and less dormant but gave lower percentage of germination than tertiary and quaternary umbels.

This is attributed to the imbalance of gibberellin and cytokinin levels in seeds. Dormancy is also induced in seeds by exposing them suddenly to dry conditions after the imbibing state , while seed germination improves with a combination of GA and PEG. Incorporation of growth substances into PEG is more effective than soaking seeds in gibberellins and ethylene solution alone, either before or after osmotic priming.

## Storage Conditions

Celery seeds maintain high viability and vigor at low temperature and lower humidity. The choice of temperature range and packaging is selected based on duration of storage. For short storage, well-dried seeds can be packed in polyethylene bags and stored at 5°C. For longer storage, seeds can be packed in moisture-proof containers, such as laminated aluminum foil pouches, and then stored at subzero temperatures.

# Chapter 64

# ASPARAGUS: *ASPARAGUS OFFICINALIS* L.

## INTRODUCTION

Asparagus is an important vegetable in Europe. It is cultivated for its edible tender green shoots (spears). Immature shoots are cooked as a vegetable and also processed and canned.

### Origin and Distribution

Asparagus originated in the Mediterranean region. It is widely cultivated in Australia, China, England, France, Germany, Hungary, Indonesia, Japan, Kenya, Malaysia, Mexico, the Netherlands, Spain, Taiwan, the United States, and the West Indies.

### Morphology

Asparagus belongs to family Liliaceae and has diploid chromosome number 2n = 20, 40. It is a monocotyledonous dioecious herb. It has large fleshy tuberous roots that emerge from an underground rhizome. Leaves are narrow and dark green. Male flowers are small and yellowish green, while female flowers are somewhat large. Fruit is a berry that becomes red on maturation and contains one to three seeds. Seeds are round, flattened, and black in color. One thousand seeds weigh about 20 to 40 g.

## SEED STORAGE

Seeds are used in propagation of asparagus. They are also used in crop improvement. Asparagus seeds show orthodox storage behavior. Higher storage temperatures and relative humidity affect seed longevity.

### Seed Collections

Asparagus is a cool-season crop that grows well in rich humus soil. Seeds used for propagation are sown in the nursery and later transplanted to the field. Normal cultural practices are adopted during growing. It is a crosspollinated crop, and insects mainly bring about pollination. About 1,000 m of isolation distance is maintained between two genotypes. Off-type plants removed regularly to maintain genetic purity. Ripe berries are harvested, and seeds are extracted, washed, cleaned, dried, and packed for storage.

### Seed Germination

Low germination is a common phenomenon in asparagus seeds , which show slight dormancy that can be removed by after-ripening for three weeks of storage. However, longer storage, for three months, is better for higher germination. For higher viability and vigor, seeds extracted from red ripe berries are superior over mature bronze berries. Similarly, seeds from a late-harvested crop show low viability.

*Seed size.* Seed size also determines the rate of germination. Large and medium-sized seeds gave a higher coefficient of germination and better seedling growth.

### Storage Conditions

Seed longevity predominantly depends on storage conditions. However, the initial vigor of seeds contributes to a certain extent. Seeds stored under improper conditions, such as high relative humidity and high temperature, deteriorate faster and lose their viability and vigor rapidly. Factors that promote delay of germination during storage enhance the seed longevity. Komoti reported that drying induced secondary dormancy in seeds.

Similarly, a higher level of CO2 concentration in storage induces secondary dormancy. Seed storage at lower temperature (4°C) maintains higher viability and vigor than storage at a higher temperature (20°C). Well-dried seeds stored at a lower temperature have better storage life.

# CUCURBITS

# Chapter 65

# CUCUMBER: *CUCUMIS SATIVUS* L.

## INTRODUCTION

Cucumber is cultivated in tropical and subtropical regions for its edible fruits. Tender fruits are eaten raw in salad, cooked as a vegetable, and also pickled. They contain carbohydrates and are a rich source of vitamin C.

### Origin and Distribution

Cucumber originated in India, and large genetic variability is observed in different parts of India. It is grown in Bulgaria, China, Egypt, Greece, India, Indonesia, Japan, Malaysia, Mexico, the Netherlands, Poland, Russia, Spain, Turkey, and the United States.

### Morphology

Cucumber belongs to Cucurbitaceae family and has diploid chromosome number 2n = 14. It is a monoecious, dicotyledonous annual plant. Vines are long and trailing with stout, hairy stems. Leaves are triangular, have three to five lobes, and are cordate. Flowers are yellow; male flowers are borne in clusters and outnumber female flowers. Fruit is a pepo and varies in size and shape, from globular to oblong, and is many seeded. The skin color is whitish green to dark green, turning yellow or brown on maturity. Flesh is pale green with no central cavity. The bitterness of fruit and foliage is owing to the terepene compound cucurbitacin, which gives resistance to certain insect pests. Seeds are white and flat. One thousand seeds weigh 25 to 33 g.

## SEED STORAGE

In cucumber, seeds are used for propagation and in breeding programs. Seeds can be stored until the next growing season or for a very long period for genetic conservation. Cucumber seeds exhibit orthodox storage behavior. They can withstand removal of moisture and can be stored at lower temperatures. High seed viability and vigor are retained at lower temperatures and lower moisture content. Nakamura noted that cucumber seeds retain their viability for a fairly long period. They lose viability completely by 38 months of storage. Germinating power and yield reduce steadily in crops raised from seeds stored under ambient conditions.

Crop yield is lower in plants raised from older seeds than in those from fresher seeds. On proper seed storage, low deterioration occurs. Such seeds give more productive plants than fresh ones. Boos ( noticed that cucumber plants raised from two- to three-yearold seeds are more productive than plants raised from one-year-old seeds. In fresh seeds, catalase activity is lower than in stored seeds, suggesting that the vigor of freshly harvested seeds is lower than that of stored seeds. Likewise, there was a delay in ethylene production in fresh seed germination. Seed vigor improves in fresh seeds following after-ripening at 75°C for 24 h. Chemical composition of seeds varies with storage period. Sugars and vitamin C tend to increase with longer storage.

### Seed Collections

Cucumbers are propagated by seeds. High seed viability and vigor are essential for high plant populations in the field and, in turn, for higher crop production. Cucumber grows well in warm climatic conditions. Seed quality improves with growing plants in artificial light, limiting the number of fruits per plant, and pollinating artificially with bees. Seed yield increases and fruit size and number of fruits per plant decrease with an increase in plant population.

Cucumber prefers deep, fertile, welldrained soils. Seeds are directly sown in the field on ridges or on hills. Plants are regularly irrigated, properly fertilized, and protected against pests and diseases. Plants are cross-pollinated mainly by bees, and an isolation distance of 1,600 m is maintained to prevent genetic contamination. Male flowers open earlier than female flowers; the latter are favored by short-day conditions and low temperatures. Plants are to be examined for off types, especially during the flowering and fruiting stages. Ripe fruits are collected when they turn pale yellow or golden in color (see Figure.1). Seeds are separated by either mechanical or chemical means. The pulp along with the seeds is fermented. Seeds that sink to the bottom are selected, and floating ones are discarded. Selected seeds are dried, cleaned, and stored.

*Fruit maturity.* Fruit maturity coincides with seed maturity. Seed maturity determines seed quality for sowing and storage. In cucumber, seed quality is poor in seeds obtained from immature fruits (20 days after anthesis), even after subjecting fruits to postharvest ripening. Seeds harvested at 30 days after anthesis need postharvest ripening of 7 to 15 days for better quality. The best seed quality is obtained 40 days after anthesis. Further delay affects seed quality by excessive respiration and field damages. The lipase activity and vitamin C content increase during seed germination. In seedlings raised from seeds extracted from mature fruits, lipase activity, ascorbic acid content, and other biochemical indices were at higher levels than in seedlings raised from seeds of immature fruits. Mature seeds retain viability better than seeds of immature fruits.

*Figure.1. Cucumbers Ready for Harvest*

## Seed Germination

Seed germination is epigeal and completes in five to seven days. It is influenced by fruit maturity at harvest and duration of fermentation, in seed extraction, seed storage time, seed placement, and germination temperatures. Solanki and Seth reported that a constant temperature of 30°C is ideal for germination. The germination process is completed in the shortest time at higher temperatures. Certain cucumber cultivars germinate more quickly than others, and heritability for speed of germination is very low (0.15 to 0.20 percent).

*Seed dormancy.* Shifriss and George observed dormancy in seeds of cultivated, wild, and hybrid cucumbers that can be overcome by afterripening for one month. Seeds stored for six months at 15 and 20°C showed improved germination. Light inhibits germination, and longer exposure at 20°C induces secondary dormancy.

*Seed placement.* Seeds placed at 45° angles with the embryos pointed upward give high germination percentage and best seedling growth. This was attributed to better hypocotyl emergence from the soil, rapid cotyledon expansion, and earlier differentiation of leaves. However, MacNeill and Hall observed that sowing seeds in the flat position gave higher germination at a constant temperature of 26.7°C.

*Chemicals.* Seed germination improves with the application of certain growth substances. Seed treatment with GA 4/7 plus ethephon or insulin or insulin-like factors I and II enhances the germination of cucumber seeds. Further, seed quality is improved by treating seeds with succinic acid.

*Storage of pregerminated seeds.* Pregerminated seeds can be stored for ten days in a cold-air medium. Seed viability and vigor decrease on longer storage, which is attributed to the degradation of functional structures, an increase in fatty acids content, and enzymatic degradation.

## Storage Conditions

Seed viability decreases rapidly at higher storage temperature and relative humidity. Cucumber seeds fail to germinate at 20 to 30°C and 82 percent RH after three months of storage. For shorter seed storage, the optimum conditions are 25°C and 45 percent RH. Zaitseva reported that seeds stored at a constant temperature of 16 or 25°C or at varying temperatures of –7 and 20°C maintain viability for seven years. Seed viability decreases rapidly at 20°C and 80 percent RH. The optimum conditions for storage are temperatures close to zero and relative humidity not exceeding 70 percent. At lower temperature and lower humidity, such as 3°C and 38 percent RH, seeds retain 80 percent germination after ten years of storage. Further, none of the seeds germinated after 13 years of storage. Seed moisture should be low for ultralow-temperature storage. Seeds with 5 or 6 percent moisture stored at –196°C retained germination ability for two years, and germination was affected when seed moisture increased to 9 to 10 percent. The safe moisture level is 1.02 to 4.08 percent for ultradry storage. Seeds with this moisture range retain high viability and vigor; those with moisture below this range are dam aged by extreme drying, and those with moisture above this range showed reduced storability.

*Seed packaging.* Various kinds of packaging materials are employed in seed storage. The ideal material should protect seeds from high moisture, withstand low temperatures, and preserve viability for longer periods. Sealed airtight containers are better for storage than unsealed containers. Cooper noted that sealed polyethylene film of 0.125 and 0.25 mm thickness is an excellent moisture-resistant package for dry cucumber seeds. Thin polyethylene film of 0.037 to 0.05mmwas useful for a short-term marketing package, but not for long-term storage.

*Storage fungi.* Seed quality is affected during rainy or humid conditions. Fungi invade seeds, especially in open storage, and reduce seed viability and vigor. Bujdoso recommends seed treatment with Ceresan, thiram, or captan to protect seeds from storage fungi and also to maintain high germination.

## Invigoration of Stored Seeds

Seed germination and seedling vigor decrease with an increase in storage period. Use of certain chemical treatments during midstorage revives vigor. Solanki and Joshi reported that seed treatment with potassium orthophosphate (3 percent) improves the seed germination of four-year-old cucumber seeds. Seed coating with calcium chloride (40 millimolar [mM]) had a similar effect.

# Chapter 66

# MUSKMELON: *CUCUMIS MELO* L.

## INTRODUCTION

Muskmelon is widely grown in the tropics. It is also known as sweet melon, cantaloupe, or honeydew melon. It is cultivated for its edible fruits, which have a sweet and musky flavor. Fruits are commonly used for dessert purposes. They are rich in sugar and vitamins A and C. Seeds can also be eaten after removing the hard seed coat.

### Origin and Distribution

Muskmelon originated in tropical Africa. Subsequently, its cultivation spread to other tropical regions. It is largely grown in China, Egypt, France, India, Indonesia, Iran, Italy, Japan, Korea, Malaysia, Mexico, Morocco, the Philippines, Romania, Spain, Turkey, and the United States.

### Morphology

Muskmelon belongs to family Cucurbitaceae and has diploid chromosome number 2n = 24. The plant is cultivated as an annual. It has long trailing vines, with leaves that are dark green, alternate, ovate, and three to seven lobed. Flowers are monoecious or andromonoecious. Pistillate flowers are borne singly, and staminate ones in clusters. Fruit is a fleshy pepo and varies in size, shape, and color. Fruits are globular or oblong, smooth or furrowed, green, yellow, or yellow-brown; flesh is green, pink, or yellow and has a musky scented flavor. Seeds are white or cream, black, or red-brown; flattened; and number 400 to 600 per fruit. One thousand seeds weigh about 5 to 25 g.

## SEED STORAGE

Muskmelon seeds show orthodox storage behavior and withstand desiccation and chilling. Seeds remain viable for a long period under dry, cool conditions. High seed germination and vigor are retained for two years under ambient conditions. Seed storability varies in different cultivars. Some cultivars are more sensitive to aging than others. Low-vigor seeds have higher respiratory quotients.

### Seed Collections

High-quality seeds with high viability and vigor are essential for high crop production. Crops grow well in sunny locations and during the warm season. Fruits produced under humid and rainy conditions are of inferior quality and get invaded by several diseases. Muskmelon requires deep, fertile, well-drained soil. High phosphorus content in the soil improves fruit quality and yield. Seeds are sown directly on ridges in rows or raised beds.

Proper spacing, adequate nutrition, timely irrigation, and plant protection measures are followed. Muskmelon is compatible with other cucurbits and is generally pollinated by bees. To protect against genetic contamination, an isolation distance of 1,000 m is maintained. Off-type plants are removed, especially during the flowering and fruiting stages. Staminate flowers open early and outnumber pistillate flowers. Fruit color changes from green to yellow or white on maturity, and ripe ones are plucked for seed purposes. Seeds are separated from the pulp by washing and then are dried, cleaned, and packed for storage.

*Fruit maturity.* Seeds from fully matured fruits maintain high seed quality during storage. Fresh muskmelon seeds show dormancy, and an afterripening period is required at all stages of seed development. Seeds are fully mature 60 days after anthesis. Seed germination increases in fully matured seeds after one year of storage at 20°C.

After-ripening improves seed quality and the yield of plants raised from the seeds. During fruit ripening, the quantity of oligosaccharides and vitamin C content increase and monosaccharides, total acidity, and pectin content decrease. Only fruits 35 days old or older are suitable for after-ripening.

*Fruit position.* The position and number of fruits per plant influence seed quality. An increase in fruit number is accompanied by a decrease in fruit weight, number of seeds per fruit, seed weight, and germination percentage.

## Seed Germination

Seed germination is epigeal and completes in five days. Germination improves with pulp fermentation of 24 h. It was highest in fruits harvested at full maturity, as compared to those in the full-slip or half-slip stage. Seeds from the proximal and middle portions of the fruit germinate better than seeds from the distal end. Seed germination and emergence hasten with storage up to 15 months and then decline, which is attributed to changes in lipase-1-peroxidase activity. Osmotic priming improves seedling emergence, whereas it has a deleterious effect on storage life of muskmelon seeds.

## Storage Conditions

*Seed moisture.* Low seed moisture increases the storage life of seeds. Moisture content of 6.6 percent or less is ideal for long storage life of muskmelon seeds. In ultradry storage, extreme drying damages muskmelon seeds, and at a higher level of seed moisture, the storability decreases, especially at room temperatures.

*Temperature.* Seeds lose their viability rapidly under high temperatures, especially if they have high moisture content and are kept in open storage. Seeds maintain viability for two years at room temperatures, whereas the storability extends to eight years by storing them in polyethylene bags and laminated aluminum foil pouches at low (5°C) and subzero temperatures (–20°C) (Doijode, unpublished data).

*Oxygen.* A high level of oxygen reduces seed viability and vigor. Seeds stored in nitrogen atmosphere under ambient conditions exhibit higher germination percentage after six years of storage.

*Storage fungi.* Seed quality is affected by fungal infection. Fungal activity is higher at high seed moisture and high humidity. Some of the fungi that associate with muskmelon seeds are *Botrytis cinerca, Cephalosporium* spp., *Cladosporium herbarium, Fusarium moniliforme, Mucor* spp., *Penicillium oxalicum,* and *Trichothecium* spp. Heat treatment at 45°C for 10 min eliminates several fungi, without any adverse effect on seed germination. Also, fungicides such as captan and carbendazim are effective in controlling storage fungi.

# Chapter 67

# WATERMELON: *CITRULLUS LANATUS* (THUNB.) MANSF.

## INTRODUCTION

Watermelon is a popular dessert fruit in the tropics and subtropics, especially in the dry regions of the world. It is grown for the edible juicy fruit. The sweet pulp, which is rich in sugars and amino acids, is eaten raw. The juice is canned and also used in the preparation of syrup. Seeds, which are rich in edible oil, are roasted with salt and eaten.

### Origin and Distribution

Watermelon is native to south-central Africa. It is cultivated throughout the tropics and subtropics. It is grown in Brazil, China, Egypt, Greece, India, Iran, Iraq, Italy, Japan, Korea, Mexico, Morocco, Spain, Tunisia, Turkey, and the United States.

### Morphology

Watermelon belongs to the Cucurbitaceae family and has diploid chromosome number 2n = 22. Watermelon is an annual dicotyledonous plant having vigorous spreading vines with a deep root system; thus, it can survive under relatively dry conditions. Leaves are pinnately lobed, number three to four pairs, and have hairy tendrils. Flowers are unisexual, single, axillary, light yellow, small, and less showy compared to other flowers of cucurbits. Male flowers exceed the number of female flowers. Fruit is large, oblong, globular, ellipsoidal, or spherical in shape, with a thick and fragile rind.

Fruit color is green, or cream and striped. Flesh is juicy and pink to red, yellow, white or greenish white, and the central cavity is absent. Seeds are numerous, about 200 to 500 per fruit; brown, black, red, white, greenish, or yellow; and rich in carbohydrates, fats, and proteins. One thousand seeds weigh about 110 to 115 g.

## SEED STORAGE

Watermelon is exclusively propagated by seeds. Seeds are also used in hybridization programs and in the conservation of genetic diversity. Seeds show orthodox storage behavior and are fairly good storers under ambient conditions. Seeds maintain their viability for a fairly long period. In a long-term study, seeds of several species of watermelon stored for 43 to 60 years at the National Seed Storage Laboratory, Fort Collins, Colorado, lost only 51 percent of their viability.

### Seed Collections

Watermelon is a warm-season crop. It requires a hot, dry climate, plenty of sunshine, and a long growing season. It grows well in deep, fertile, welldrained, sandy loam to loam soils. It is propagated by seeds. Seeds are sown directly in the field, either on hills or on ridges. Crops are suitably spaced, fertilized, watered, and protected against pests and diseases for better crop growth and seed production. Watermelon is a cross-pollinated crop, and pollination is mainly brought about by honeybees. Male flowers appear first, followed by pistillate flowers. An isolation distance of 1,000 m is provided for maintaining genetic purity. Off-type plants showing undesirable vegetative and fruit characteristics are removed, especially during the flowering and fruit development stages. Fruit takes about 80 to 110 days for maturity, and its color changes from green to pale yellow on the underside of the fruit. Seeds are located in the central areas of the fruit pulp. Seeds are removed along with pulp, washed, dried, cleaned, and processed for storage.

*Fruit maturity.* Fruit color changes on maturity and becomes light yellow, especially on the lower surface. During this stage, seeds also attain maturity and give higher germination. Watermelon fruit matures in 49 to 54 days. Normally, fruits are left on the vines for a comparable period after the vines become dry. In cucurbits, the seed coat completes its growth earlier than the embryo, and this may play an important role in inhibiting the germination of immature seeds. Seed coat removal increases germination in immature seeds.

*Fruit storage.* Watermelon fruits can be stored for relatively long periods. Seeds from stored fruits also maintain seed quality for a long period. Bankole reported that high seed germination could be maintained for four months when fruits are stored under ambient conditions, but germination decreases on prolonged storage. Nerson and Paris observed that storage of mature fruits up to 48 months did not affect seed germination, but germination was affected in seeds of immature fruits.

## Seed Germination

Watermelon seeds possess a hard seed coat and thus take longer to germinate. Seed germination is epigeal and slow at lower temperatures. It improves with simple washing of seeds with water and scarification of the seed coat. Kolev and Boyadzhiev reported that seed germination and storage life improved by treating the freshly extracted seeds with 2 percent hydrochloric acid or sodium hydroxide. Seed germinates readily on full imbibitions. Very dry or excess seed moisture affects the germination process. Seed germination is suppressed at minimum and maximum levels of 5 and 29 percent moisture. Seed germination was higher under darkness, and intermittent exposure of seeds to red or far-red light affects the germination process.

## Storage Conditions

*Seed moisture.* High seed moisture is injurious to seed viability. Moisture content of 5 to 7 percent is safe for long-term storage. Excessive removal of moisture damages cell structure and affects seed quality. Ji, Guo, and Ye recommended moisture content of 1.25 to 4.26 percent for ultradry storage. Seed moisture less than 1.25 percent damages the cell structure, thereby reducing storability.

*Seed packaging.* A suitable container is required to maintain a desirable moisture level in seeds. Seeds can be stored in cloth bags, paper bags, glass containers, and laminated aluminum foil pouches, or aluminum cans, depending upon storage period and conditions. Polyethylene bags and laminated aluminum foil pouches are effective in maintaining high viability at lower temperatures. Polyethylene bags are also cheaper, transparent, and sealable. Watermelon seeds stored in polyethylene bags at 2°C and 65 percent RH retain higher viability than seeds stored in paper bags. In a hermetically sealed box, seed viability was preserved for 15 to 23 years, and seeds packed in paper bags exhibited lower germination and more abnormal seedlings after four to five years of storage.

*Temperature.* High temperature promotes seed deterioration. Seeds stored at room temperatures remain viable for only two years, as compared to 15 years in those stored at 5 and –20°C.

*Air.* Watermelon seed quality decreases during open storage under ambient conditions. A high level of oxygen promotes the respiration process. Seeds stored in nitrogen remain viable for seven years under ambient conditions.

# Chapter 68

# PUMPKIN: *CUCURBITA MOSCHATA* (DUCHESNE) POIR.

## INTRODUCTION

Pumpkin is an important cucurbitaceous vegetable in arid regions of the tropics. It is cultivated for its edible fruits, which have a long storage life under ambient conditions. Ripe fruits are cooked as a vegetable and also are fed to cattle. Pumpkin is rich in carbohydrates, potassium, and vitamins A and C.

### Origin and Distribution

Pumpkin originated in the arid regions of Central and South America. It is grown largely in the tropics. Some major pumpkin-growing countries are Argentina, China, Egypt, India, Indonesia, Iran, Italy, Japan, Malaysia, Mexico, Peru, the Philippines, South Africa, Spain, Turkey, and the United States.

### Morphology

Pumpkin belongs to the Cucurbitaceae family. It is a dicotyledonous trailing plant having chromosome number 2n = 40. Leaves are simple, large, alternate, and deeply lobed. Plant is monoecious and bears a single yellow staminate flower on a long stem, while the pistillate is borne on a shorter stem. Fruit is large and varies in size, and its flesh color ranges from yellow to orange. Seeds are brown or white with a broad rim and con- tain about 30 percent proteins and 40 percent oils. One thousand seeds weigh about 200 g.

## SEED STORAGE

Pumpkin is commercially propagated by seeds, which have greater value in genetic conservation. Seeds show orthodox storage behavior, and seed deterioration decreases with decreases in seed moisture and storage temperature.

### Seed Collections

Pumpkin is a warm-season crop; it tolerates hot, dry climatic conditions. It requires deep, fertile, sandy loam soils. Seeds are sown directly on ridges or hills in the field. Pumpkin is cross-pollinated by insects, mainly bees. Fruit shape and yield are affected by improper and inadequate pollination. Therefore, one or two beehives are placed near the seed production field. An isolation distance of 1,600 m from other cucurbits is maintained to avoid cross-fertilization. Roguing of off-type plants is done during the flowering and fruiting stages. Only a few fruits per vine are retained for optimum size; otherwise, fruit weight will be reduced. Heavy fruits yield bold and highquality seeds. Dematte reported that below average fruits give lower seed yield, but those seeds give the highest percentage of germination. Fruit takes normally four months from anthesis to maturity. Change of fruit color to yellow or straw color is the indication of maturity (see Figure 1).

*Figure 1. Pumpkins Ready for Harvest*

Seeds are removed along with the pulp from the matured fruit and subjected to fermentation in water for 24 h followed by washing in running water for 40 min.

*Fruit maturity.* Seed viability and vigor are high in seeds harvested at the mature stage. Araujo, Mantovani, and Silva reported that pumpkin seeds extracted from fruits 15 to 65 days after anthesis can be stored for seven weeks. Seed germination and seedling vigor increase with increasing fruit age and storage length.

*Fruit storage.* Pumpkin fruits can be stored for a fairly long period at room temperatures. Seed quality improves with storage of fruits. Goldbach reported that seeds of ripe fruits give high viability and vigor. Seed germination in unripe fruits improves after ripening. Further, storage of ripe and overripe fruits at 23°C and 50 percent RH for six weeks resulted in loss of seed vigor and/or germination.

## Seed Germination

Seed germination is epigeal, and seedling emergence takes less time at higher temperatures. Seed germination is better at alternate temperatures of 20/30°C for 16/8 h, respectively.

*Seed size.* Large seeds produce higher yield than small or medium-sized seeds, whereas there is no relationship between seed size at planting and seedling emergence.

*Chemicals.* Seed germination in pumpkin improves by soaking seeds in an ammonium chloride solution (400 ppm) for 6 h.

## Storage Conditions

Pumpkin seeds are viable for four years under ambient conditions, whereas high germination is retained for only two years. Seed viability and vigor are affected by higher storage temperatures and relative humidity. Zink and DeMendonca preserved pumpkin seeds for 18 months by storing air-dried seeds in cloth bags under ambient conditions. Moisture content of 1.79 to 4.07 percent is safe for ultradry storage of pumpkin seeds. Further, reducing the moisture damages cell constituents, and higher moisture reduces seed storability. Seed storage at 5 and –20°C maintains high viability and vigor for 15 years. The loss of viability is associated with greater leaching of electrolytes, soluble sugars, and free amino acids and a decrease in dehydrogenase activity in seeds.

# Chapter 69

# MARROW: *CUCURBITA PEPO* L.

## INTRODUCTION

Marrow is a popular cucurbit in the tropics and subtropics. It is also known as summer squash or vegetable marrow. It is cultivated for its edible fruits. Immature fruits are boiled or roasted, and mature fruits are cooked as a vegetable. It is a good source of carbohydrates and vitamin A.

### Origin and Distribution

Marrow originated in South America. Large genetic diversity exists in Mexico. It is cultivated in Argentina, Brazil, India, Indonesia, Malaysia, Mexico, the Philippines, and the United States.

## SEED STORAGE

Marrow seeds exhibit orthodox storage behavior. These seeds can be dried to a very low moisture content, and storing them at chilling temperatures can extend longevity. Seeds are viable for two to three years under ambient conditions and need cool, dry conditions for better storage.

### Morphology

Marrow belongs to family Cucurbitaceae and has chromosome number 2n = 40. Marrow is a monoecious annual that produces both bushy and longrunning plants. Stems are rough with large, deeply lobed leaves. Flowers are bright yellow to orange yellow, single, and showy. Fruits differ in shape, size, and color and have a hard or soft shell. Fruits contain a large number of seeds that are easily separable from the pulp. Flesh is yellow or white. Seeds are brown, white, and large or small with a rim. One thousand seeds weigh 300 to 310 g.

## Seed Collections

Marrow is a warm- and dry-season crop that withstands high temperatures. It grows well in well-drained, humus soil. It is propagated by seeds. Seeds are sown directly in the field, either on ridges or on hills. Plants are provided with adequate nutrients and watered regularly for better crop growth. Marrow is a cross-pollinated crop, and insects bring about pollination. Male flowers outnumber and appear earlier than female flowers. An isolation distance of 1,600mis maintained to protect from cross-pollination with other cucurbits. Plants are inspected during the flowering and fruiting stages for various morphological characteristics, and off-type plants are re moved. Fruit color changes on maturity. Dry fruits are collected, and seeds are removed from the pulp. Seeds are washed thoroughly, dried, cleaned, and packed for storage.

*Fruit maturity.* Seed germination is higher in seeds of matured fruits. Marrow seeds reach mass maturity 61 to 63 days after anthesis and moisture content decreases to 40 to 48 percent. Nerson and Paris reported that seeds extracted from immature fruits (28 days after anthesis) did not germinate. Seeds harvested 24 to 31 days after mass maturity gave higher seed longevity.

*Fruit storage.* In cucurbits, fruit storage is a common practice used to improve seed quality. Storage of ripe fruits for three weeks and of immature fruits for 12 weeks at room temperature eliminates dormancy. The mature marrow fruits can be stored for 48 months with germination unaffected.

## Seed Germination

Seed germination is epigeal and higher at alternate temperatures of 20/30°C for 16/8 h or at a constant temperature of 30°C. Seeds take one week to germinate under dark conditions. Gomaa reported that seed placement at a 45° angle, with the embryo end upward, gives a higher percentage of germination and better seedling growth.

*Seed dormancy.* Marrow seeds exhibit dormancy to a certain extent. Seed dormancy is eliminated by after-ripening of seeds for one to two months (. Decoating of seeds also overcomes the dormancy and gives 100 percent germination shortly after mass maturity.

## Storage Conditions

Seed deterioration decreases at a lower level of seed moisture and at a lower temperature. Demir reported that seed longevity improves with a decrease in seed moisture and temperature. Seed moisture of about 6.5 percent gives highest seed storability, and any subsequent increase in moisture reduces seed longevity. Root growth is also adversely affected at higher seed moisture. Thus, marrow seeds are dried to a lower moisture content (5 to 7 percent) and suitably packed in moisture-proof containers, such as laminated aluminum foil pouches, and are preferably kept at –20°C for long-term storage. However, seeds can be packed in polyethylene bags and stored at 5°C for relatively longer storage.

## Invigoration of Stored Seeds

Seed priming improves the rate of seedling emergence, produces uniform seedlings, and gives a higher germination percentage, irrespective of the osmotic agent used or the duration of treatment.

# Chapter 70

# RIDGE GOURD: *LUFFA ACUTANGULA* (L.) ROXB.

## INTRODUCTION

Ridge gourd is also called angled gourd, angled loofah, Chinese okra, and silky gourd. It is cultivated in the tropics for its tender edible fruits. Fruits are cooked as a vegetable. Seeds are rich in protein and oil.

### Origin and Distribution

Ridge gourd originated in India. Wide genetic variation for various morphological fruits is observed in different parts of India. It is cultivated in India, Indonesia, Malaysia, Myanmar, the Philippines, Sri Lanka, and Taiwan.

### Morphology

Ridge gourd belongs to family Cucurbitaceae and has chromosome number 2n = 26. The plant is a dicotyledonous, monoecious, annual climber with tendrils. Leaves are pale green, simple, and ovate. Staminate flowers are many and arranged in a raceme, while pistillate flowers are solitary and fragrant. Sometimes hermaphrodite types also exist. Fruit is club shaped and angled, with ten prominent ribs, and many seeded. It becomes hard, bitter, and inedible on maturity. Seeds are black, flattened, and wrinkled and contain a high amount of oil. One thousand seeds weigh about 150 to 170 g.

## SEED STORAGE

Ridge gourd seeds show orthodox storage behavior. Seed storage is enhanced by drying seeds and placing them under cooler conditions. Seeds are viable for two to three years under room temperatures. Higher storage temperature and relative humidity reduce storage life.

### Seed Collections

Ridge gourd is a warm-season crop. It grows well in hot, humid climates. It prefers deep, humus, sandy loam soil. Seeds are sown directly in the field, either on hills or on ridges. Sufficient farmyard manure and chemical fertile izers are applied based on soil conditions. Timely irrigation is given, especially during the dry season at the flowering and fruit development stages. It is a cross-pollinated crop, and pollination is brought about by insects. An isolation distance of 800 m is provided to prevent genetic contamination.

Plants are checked for various morphological characteristics during the flowering and fruiting stages, and off types are removed. Fruit matures in four to five months and changes color on maturity. Dry fruits are collected, and seeds are separated.

### Seed Germination

Seed germination is epigeal and better at alternate temperatures of 20/30°C for 16/8 h, respectively. Seed germination is rather slow and uneven and takes longer to complete, due to the hard seed coat. Fursa and Gvozdeva  noted that moist stratification of seeds at 30 to 35°C for three days and warm stratification at 40 to 50°C for two or three weeks gives higher percentage of germination.

### Storage Conditions

High seed viability and vigor are maintained under proper storage conditions. These well-stored seeds germinate within five days, give better seedling growth and development, induce flowering early, and give a higher yield than low-vigor seeds. Ridge gourd seeds were successfully preserved in aluminum foil pouches at 20 to 33°C and 67 to 86 percent RH for ten months, contrary to many reports claiming viability losses with packets. Further, well-dried seeds (5 to 7 percent mc) packed in polyethylene bags and stored at 5°C maintain high viability and vigor during storage. For long-term storage, it is preferable to pack low-moisture seeds in laminated aluminum pouches and to store them at subzero temperatures (–20°C).

# Chapter 71

# SPONGE GOURD: *LUFFA CYLINDRICA* (L.) ROEM

## INTRODUCTION

Sponge gourd is also called dishcloth gourd, hechima, rag gourd, smooth loofah, and vegetable sponge. Sponge gourd is cultivated both for its use as a vegetable and for fiber purposes. The tender fruits of nonbitter types are cooked as a vegetable, and fiber is extracted from mature fruits. It is a good source of carbohydrates, calcium, phosphorus, iron, and vitamins A and C.

### Origin and Distribution

Sponge gourd is native to India. Subsequently, it spread to Southeast Asian countries and other tropical regions. It is cultivated in Argentina, Brazil, India, Indonesia, Japan, Malaysia, the Philippines, and Sri Lanka.

### Morphology

Sponge gourd belongs to family Cucurbitaceae and has diploid chromosome number 2n = 26. It is a dicotyledonous, vigorous, annual climbing vine. Stems are rough and five angled. Leaves are dark green with five to seven lobes. Plant is monoecious, and male and female flowers are borne separately on the same vine. Male flowers appear in clusters and number from 4 to 20. Female flowers are single and axillary. Fruit is cylindrical with light stripes and many seeded. Seeds are black, flat, and smooth, with a narrow wing, and about 10 to15 mm long. They are rich in oil. One thousand seeds weigh about 90 to 100 g.

## SEED STORAGE

Sponge gourd is primarily raised through seeds, which show orthodox storage behavior. Seeds are fairly good storers between growing seasons under ambient conditions.

### Seed Collections

Sponge gourd requires a hot climate during growth. Excessive rainfall during flowering and fruiting affects yield as well as seed quality. It grows well in fertile, well-drained soil. Seeds are sown on ridges or on hills in the field. Plants are provided with adequate nutrients, timely irrigation, and plant protection measures. Plants are cross-pollinated by insects. A 1,000 m isolation distance is maintained between a sponge gourd field and other cucurbits.

Fruits become yellow on maturity. Dry mature fruits are harvested, and the seeds are extracted, washed, dried, and suitably packed for storage.

### Seed Germination

Seed germination is epigeal and higher at a constant temperature of 30°C or following scarification of seeds kept in moist sand at 30 to 35°C for three days , whereas Singh and Mathur  obtained higher seed germination through warm stratification (40 to 50°C) for two to three weeks. Higher temperature during storage causes progressive destruction of seeds.

In sponge gourd, seedling growth is better with large seeds, and germination is unaffected by seed size but is maximum in black seeds as compared to other colored seeds.

### Storage Conditions

Dry, cool storage conditions are ideal for long storage of seeds. Wellmatured seeds are dried to 6 to 7 percent moisture, packed in polyethylene bags, and stored at a low temperature (5°C). Seeds are packed in moistureproof containers, such as laminated aluminum foil pouches, and stored at subzero temperature (–20°C) for long storage.

# Chapter 72

# BITTER GOURD: *MOMORDICA CHARANTIA* L.

## INTRODUCTION

Bitter gourd is an important vegetable crop in Southeast Asia. It is mainly cultivated for its bitter unripe fruits. It is also known as alligator pear, balsam pear, bitter cucumber, bitter melon, ku gua, and foo gwa. Immature fruits are cooked as a vegetable. Leaves and fruits are used in medicine. It is valued for its high calcium, phosphorus, iron, and vitamin C contents. Dipping the fruits in salt water reduces the bitterness.

### Origin and Distribution

Bitter gourd originated in India. Large genetic diversity is found in different states of India. It is a popular vegetable crop in Brazil, China, India, Indonesia, Malaysia, the Philippines, Singapore, and Sri Lanka.

### Morphology

Bitter gourd belongs to the Cucurbitaceae family and has chromosome number 2n = 22. It is an annual, monoecious, dicotyledonous climbing vine. Stems are slender with palmate, deeply lobed, simple leaves. Flowers are small, yellow, and borne singly at the axil. Fruit is green, pear-shaped, with a smooth or pointed protrusion on the surface, and many seeded. It bursts on maturity, showing red arils. Seeds are brown, 1 to 1.5 cm long, and oval with flattened arils. One thousand seeds weigh 60 to 170 grams.

## SEED STORAGE

Seeds are used in propagation and in crop improvement through hybridization. Seeds are a valuable tool for long-term conservation of germplasm. Bitter gourd seeds show orthodox storage behavior; they can tolerate desiccation and preserve their viability at lower temperatures.

### Seed Collections

Bitter gourd grows well in hot, humid areas. It prefers fertile, welldrained soil that is rich in organic matter. It is propagated by seeds and sown directly in the field on ridges or hills. Plants are fertilized and protected against pests and diseases. Irrigation is provided during the dry season. It is a cross-pollinated crop, and different insects bring about pollination. Crosspollination increases the percentage of large seeds in the fruit. An isolation distance of 1,600 m is provided to maintain genetic purity. Plants are checked for various morphological characteristics, especially during the flowering and fruiting stages, and abnormal ones are discarded. Fruit turns yellow or orange on ripening, and then it splits, exposing seeds. Ripe fruits are collected, and seeds are removed, washed thoroughly, dried, cleaned, and processed for storage.

*Fruit maturity.* Seeds continue to mature in the fruit, even if the fruit is harvested before ripening and stored under ambient conditions. Harvesting of fruits three days before full ripening and storing them under ambient conditions for four days gave high germination percentage and vigor; this also avoided field losses due to bird or rain damage.

*Fruit size.* Bitter gourd fruit size varies in different cultivars and depends on growing conditions. Seeds from very long fruits (30 cm) give higher seed germination and fruit quality, whereas germination is poor in seeds from small fruits (15 cm).

### Seed Germination

Seed germination is epigeal and higher at a constant temperature of 30°C or at alternate temperatures of 20/30°C for 16/8 h. Seed germination is rapid and greater in water-soaked seeds. Boron application or rupturing the hard seed coat can improve germination.

*Seed dormancy.* Bitter gourd seeds show seed coat dormancy and endogenous dormancy. Fresh seeds do not germinate readily, and they require afterripening for one month. Application of potassium nitrate (1 percent) eliminates the dormancy and promotes seed germination.

*Light.* Light inhibits the germination process in bitter gourd. However, daylight along with gibberellic acid (50 ppm) application enhances seed germination. Further, seed exposure to red light improves the germination, unlike exposure to green and yellow light, which showed an inhibitory effect.

*Growth substances.* Seeds soaked in a low concentration of gibberellic acid (100 to 200 ppm) and indoleacetic acid (100 to 200 ppm) showed im proved seed germination and seedling emergence.

*Seed position.* Position of seeds placed in soil affects emergence of seedlings. Seeds placed vertically give better seedling emergence and show higher seedling vigo.

## Storage Conditions

Bitter gourd seeds remain viable for two to three years at room temperatures. Seeds with 5 to 7 percent moisture maintain high viability at low temperatures. Seeds can be stored at 5°C in polyethylene bags, but for long storage, seeds are packed in laminated aluminum foil pouches and stored at –20°C. Seed deterioration is minimal under cool, dry storage conditions. The loss of seed vigor precedes a decrease in viability and is manifested in a greater leaching of metabolites from seeds. Seed germination was reduced to 62 percent on aging of seeds, and this was positively correlated with the leaching of soluble sugars and amino acids (see Table.1).

*Table.1. Effect of High Temperature (42°C) and High Humidity (95 percent) on Seed Quality in Bitter Gourd*

| Storage Period (days) | Germination (%) | Vigor Index | Electrical Conductivity (mmhos) | Soluble Sugars ($mg \cdot g^{-1}$) | Free Amino Acids ($mg \cdot g^{-1}$) |
|---|---|---|---|---|---|
| 0 | 90 | 1527 | 455 | 1.93 | 8.41 |
| 2 | 82 | 1382 | 466 | 2.04 | 8.90 |
| 4 | 72 | 1206 | 481 | 2.08 | 9.18 |
| 6 | 66 | 1097 | 532 | 2.25 | 10.18 |
| 8 | 62 | 1007 | 546 | 2.52 | 10.81 |

# Chapter 73

# BOTTLE GOURD: *LAGENARIA SICERARIA* (MOL.) STANDL.

## INTRODUCTION

Bottle gourd is an important gourd having a wide range of uses. It is largely cultivated in the tropics and subtropics for its edible fruits. It is also known as calabash gourd, trumpet gourd, white-flowered gourd, and zucca melon. It is a good source of carbohydrates and calcium. Tender fruits are cooked as a vegetable, and the hard shell is used in making various types of utensils for storage of liquid and food materials. It is also used in designing certain musical instruments. The pulp and leaves have medicinal properties.

### Origin and Distribution

Bottle gourd originated in south-central Africa and probably later spread to South America, where it exhibits a certain amount of genetic variability. It is cultivated in Brazil, China, Colombia, Egypt, France, India, Indonesia, Japan, Malaysia, the Philippines, Spain, Sri Lanka, and the United States.

### Morphology

Bottle gourd belongs to family Cucurbitaceae and has chromosome number 2n = 22. It is a monoecious plant with long climbing or trailing vines. Leaves are simple, large, pale green, and hairy. Flowers are single, axillary, and white; borne on a long peduncle; fragrant; and short-lived. Fruit varies in size and shape, being flat, globular, or bottle or club shaped, and green or white in color. Rind becomes hard on maturity. Seeds are dicotyledonous, obovate, about 2 cm long, white or brown, and ridged. One thousand seeds weigh about 130 to 150 g.

# SEED STORAGE

Bottle gourd is propagated commercially through seeds. Farmers preserve seeds in viable condition between growing seasons, whereas breeders want to store them for very long periods without affecting the genetic architecture of the genotypes. Both types of storage demand proper care and suitable storage conditions. Bottle gourd seeds lose viability under ambient conditions. Viability decreases to 50 percent after 12 months, and none of the seeds germinated after 24 months of storage under ambient conditions.

## Seed Collections

Preharvest factors play an important role in seed storage. Seeds obtained from healthy, normal crops are vigorous and store for comparatively long periods. Bottle gourd gives optimum yield with moderate rainfall, sunny days, and warm climate. It grows well in sandy loam soils rich in organic matter. Seeds are sown directly in the field on ridges or mounds. Optimum fertilizer and regular irrigation during the dry period produce healthy crop growth. Bottle gourd is cross-pollinated by insects, and an isolation distance of 800 m is maintained. Off-type plants are removed to safeguard genetic purity. Mature and dry fruits are harvested, and the seeds are removed, cleaned, dried, and stored.

## Seed Germination

Fresh bottle gourd seeds show a certain amount of dormancy. It takes two to three months of after-ripening to eliminate the dorman. Seed scarification and/or sowing at alternate temperatures of 20/30°C for 16/8 h enhances germination. Sandy soil and a constant temperature of 30°C is ideal for optimum seed germination , and germination is nil at 35°C. Seeds placed vertically with the embryo end facing upward give early and higher seedling emergence. Seed germination improves with soaking seeds in gibberellic acid at 200 pp and in succinic acid (600 ppm). Nath and Mathur observed that seeds exposed to red light gave higher germination than those exposed to blue or brown light.

## Storage Conditions

Storing dry seeds in a cooler environment extends their longevity. Seeds packed in laminated aluminum foil pouches and stored at 20 to 33°C and 67 to 86 percent RH can be stored for ten months. Further, seeds remain viable for five years at 5 and –20°C. Seeds retain high viability when stored in polyethylene pouches at 5°C and in laminated aluminum foil pouches at –20°C . Seeds retain high germination in polyethylene bags and laminated aluminum foil pouches for ten years. On deterioration, seed germination and seedling vigor decrease. The soluble sugar and free amino acid contents are higher in leachates of aged seeds.

# ORNAMENTAL CROPS

# Chapter 74

# ROSE (*ROSA* SPP.)

## INTRODUCTION

The rose is an important commercial flower, widely appreciated and valued for its beauty, color, and fragrance. Roses are largely used as cut flowers and planted in beds and gardens; rose shrubs are also used in landscaping. The present-day cultivars are derived from the natural hybridization of Chinese and European roses. Roses are broadly grouped into hybrid tea, floribunda, polyantha, climber and rambler, miniature, and shrub roses. Some of the genotypes bear attractive colored fruits called hips. Flowers of hybrid tea and floribunda roses are largely used as cut flowers.

The rose is a perennial, spiny, deciduous plant. Leaves are alternate and pinnate. Flowers are solitary or borne in clusters, and they vary in size and color, namely, white, yellow, orange, pink, red, and lavender. Fruit is a fleshy hip that contains hairy achenes. Rose grows well in temperate conditions and is propagated by seeds, cuttings, budding, and grafting. Seeds are normally used for evolving new cultivars through hybridization, while budding is predominantly employed for propagation.

## SEED STORAGE

### Seed Germination

Preservation of high seed viability and seedling vigor is the main criterion for seed storage. In nature, rose plants set seeds poorly and in small numbers. Rose seeds do not germinate immediately after harvest due to the impervious nature of their seed coat and the presence of innate dormancy. A certain period of after-ripening is necessary before the seeds are ready to germinate. Seeds are extracted from mature fruits when their color changes from green to red for higher germination.

Crushing the hips and removing the fleshy parts improves seed germination. Seeds from older hips germinate better than those from younger ones. Hips about 15 to 20 weeks old show a higher germination percentage. Seed scarification, temperature stratification, and applications of growth substances hasten the germination process. Moist seed storage at 4°C for ten weeks overcomes the cold requirements.

Seed soaking in a zinc and hydrochloric acid mixture softens the pericarp and improves the germination percentage. Seed coat removal hastened germination by seven days at 20°C, but it is a cumbersome process. The combination of warm and cold stratification in a sequence improves the germination process in certain *Rosa* species.

Seeds of *R. canina* preserved at 26.7°C for eight weeks followed by storage at 4.4°C for 12 weeks resulted in higher germination, while seeds of *R. laevigata* did not respond to warm or cold stratification and required a long storage period for germination. Popcov and Buc obtained high germination by soaking seeds in concentrated sulfuric acid for 1 h followed by warm stratification (20 to 25°C for two to three months) and cold stratification (5 to 6°C for three to five months). Blundell and Jackson suggested soaking the seeds in concentrated sulfuric acid for 1 h and then storing them at 26°C for three months followed by storage at 2°C. Yu, Yeam, and Kim reported that cold stratification for 60 days is sufficient for higher germination of *R. multiflora* seeds. Seed treatment with growth substances promotes the germination process. Seed soaking in gibberellic acid (10,000 mg·liter –1) and benzyladenine (100 mg·liter –1), followed by warm (20°C) and cold (5°C) stratification for four weeks, yielded higher germinatio. Presowing seed treatment with vitamins B1, B2, and C increases the germination as well as seedling vigor in seeds of *R. canina.*

## Storage Conditions

Rose seeds preserve well under ambient conditions. Seed deterioration is slow in different *Rosa* species. The presence of seed dormancy is beneficial for preserving viability during storage. Rose seeds exhibit orthodox storage behavior. Seed longevity improves with a reduction in moisture and storage at low temperature. Seeds of certain *Rosa* species were preserved in dry conditions for two to four years. Seeds of *R. rugosa* had better germination after three years of storage. The percentage of germination dropped in seeds of *R. multiflora* from 72.5 to 48.4 in three years and in seeds of *R. setigera* from 53.4 to 35.6 after two years of storage. Harrington indicated that a rise in seed moisture and in storage temperature reduces the storage life of seeds. Before storage, seeds are thoroughly washed in water, and all adhering pulp material is removed. Later, seeds dried to 5 to 6 percent moisture are preferably stored at low temperature (15°C) and low humidity (10 percent RH) for safer and longer storage. Dried seeds packed in suitable moisture-proof containers, such as thick polyethylene bags and aluminum foil or cans, and stored at a low temperature can preserve high seed viability and vigor during storage.

# Chapter 75
# ORCHID

## INTRODUCTION

Orchids are the most beautiful flowers in the biosphere because of their unique shape and attractive coloring. They vary in shape, size, and color. These flowers bloom on the plants over a period of about three to four weeks. Orchids are grown in gardens and in pots and are used as cut flowers.

Many orchids of different genera and species flourish well in humid, tropical conditions. These are grown predominantly in Australia, India, Malaysia, Mexico, Myanmar, New Guinea, the Philippines, Sri Lanka, Thailand, and the United States. Orchids are either terrestrial or epiphytic, the latter being in the majority and widely distributed in forested areas. The important cultivated orchid genera are *Cymbidium, Dendrobium, Paphiopedilum, Phalaenopsis, Pleione,* and *Vanda.*

Orchid is a monocotyledonous, herbaceous, perennial plant. Flowers vary in shape, size, and color, and they are fragrant in certain species. Orchid is propagated both sexually and asexually. The former causes variation in various plant characteristics and hence is used sparingly in commercial propagation. Cuttings, offsets, or meristem cultures are used for orchid propagation.

Seeds are used in evolving new genotypes through hybridization and also in the conservation of genetic diversity in the seed bank. Orchid seeds demand aseptic conditions while raising seedlings, and in nature, germination requires the help of fungi. Each capsule contains about two to three million minute seeds, and only a small percentage (0.2 to 0.3 percent) germinate.

# SEED STORAGE

## Seed Germination

Seed germination is very low and slow in orchids. Seeds germinate in association with mycorrhizal fungi, mostly nonsporal types, such as *Rhizoctonia languinosa, R. mucoroides,* and *R. repens.* Orchid seeds lack endosperm, and they depend on other sources such as fungi for food. Orchid seeds are dependent on extraneous source for their sugar supply, particularly during germination. Seed germination is higher and faster in a sugar medium, especially the Knudson's C solution. Linden reported that immature seeds of orchid germinate better and show higher seedling growth than dried mature seeds in two mineral media:

Fasts' and Burgeff-Egl. Seedlings grow better with powdered bark of incense cedar *(Libocedrus decurrens)* or sugar pine *(Pinus lambertiana)* added to the medium. Mature seeds germinate earlier and develop protocorms. The germination percentage was higher (75 percent) in 11-month-old seeds of *Vanda roxburghi.* Embryos of terrestrial orchid seeds are cultured in a medium containing macro- and microelements, sucrose, amino acids, and vitamins and are incubated in the dark at 25°C. Seeds of several orchid genotypes germinate better in a culture medium. Additions of banana pulp, coconut milk, or tomato juice to the medium stimulate the germination process.

## Storage Conditions

Orchid seeds exhibit orthodox storage behavior. Seeds retain high viability and vigor under dry, cool conditions. Seeds lose their viability rapidly under room temperatures without desiccation. The percentage of germination is low in fresh seeds. Orchid seeds stored at 6°C showed high germination (90 percent) after ten weeks of storage, while at 45°C storage, they exhibited 40 percent seed viability in the presence of a desiccant. In the absence of a desiccant, the percentage of seed germination was reduced to 73 and 25 percent after ten weeks of storage at 6 and 45°C, respectively. Orchid seeds preserved below –70°C are supposed to retain 50 percent viability for a minimum period of two centuries. Shoushtari et al. suggested storing orchid seeds at 4°C over calcium chloride to preserve viability for 10 to 20 years. Seed storage under cryogenic conditions maintains seed viability for longer periods.

# Chapter 76

# CACTUS

## INTRODUCTION

Cacti include several genera and species used for ornamental purposes. Some of the important ones are *Astrophytum, Cephalocereus, Cerei, Cereus, Chamaecereus, Cleistocactus, Echinocereus, Echinopsis, Ferocactus, Lobivia, Nopalea, Opuntia, Rebutia,* and *Zygocactus.* Cacti are extremely hardy plants and can withstand dry conditions. They are grown in gardens, especially in rock gardens, or in pots for ornamental purposes. Plants are modified into various formations, which bear mostly short-lived flowers.

Fruit is a fleshy, depressed globose. Certain cacti plants resemble particular animals, birds, and plants, giving them a unique appearance. Large genetic diversity in cacti occurs in South America. Seeds, cuttings, offsets, bulbils, and grafts are used for propagation.

## SEED STORAGE

### Seed Germination

Seed germination and seedling emergence is slow in cacti. Seeds of *Cereus peruvianus* Ricob germinate well at 15 to 25°C. The optimum temperature for germination of *Oreocereus trollii* seeds is between 15 and 20°C, and very few seeds germinate at 30°C. Seed germination is rapid and higher at alternate temperatures of 20/30°C in the presence of light.

## Storage Conditions

Cactus seeds appear to be orthodox in storage behavior. They preserve viability and vigor under dry, cool conditions. Seeds of *Cereus peruvianus* and *Oreocereus trollii* lose their viability rapidly under room temperatures and maintain it at 2°C. The viability of *Espostoa lanata* seeds decreases with an increase in storage period. Dry seeds are packed in polyethylene bags or in laminated aluminum foil pouches and stored at low temperatures for longer maintenance of seed viability.

# Chapter 77

# GLADIOLUS (*GLADIOLUS* SPP.)

## INTRODUCTION

Gladiolus, or sword lily, is a perennial herbaceous plant with swordshaped foliage. It is predominantly used as a cut flower and planted in beds or as a herbaceous border in a garden. Of the many species of *Gladiolus, G. colvillei, G. gandavensis, G. segetum,* and *G. tristis* are cultivated for ornamental purposes. Flowers are showy, of various colors, and arranged on one side of the spike. Flowers vary in shape, size, and color and have excellent keeping quality. The inflorescence contains up to 30 flowers, and colors include red, yellow, pink, violet, orange, and purple. Corms are commonly used for propagation. The spike length greatly depends on corm size. A medium-sized corm with a high crown gives better seedling growth during propagation than the large and flat corms. Gladioli are broadly grouped into large flowered, butterfly, and miniature types. The latter ones are suitable for bouquets and are grown in smaller gardens. Seeds are used in hybridization programs for the improvement of flower quality and the incorporation of resistance to various biotic and abiotic stresses.

## SEED STORAGE

### Seed Germination

Light and temperature influence seed germination and germination rate in gladioli. A constant temperature of 20°C is congenial for higher total germination (97 percent), which takes about 16 days to complete.

## Storage Conditions

Gladiolus seeds are orthodox in storage behavior and their viability decreases at higher storage temperatures and relative humidity. Seeds remain viable for a period of one year under room temperatures. The germination percentage is reduced after 12 months of storage. The optimum storage conditions of 14°C and 26 percent RH are favorable for longer seed storage. Barton reported that high seed germination capability was retained for five days when seeds were stored in sealed containers at 5°C. Seed storage at –4°C extends the longevity to ten years or more. Seed moisture increases with repeated opening and closing of containers, thereby affecting seed viability. Thus, this should be avoided as much as possible during seed storage. However, reduction of seed moisture from 11.8 to 4.2 percent does not affect total germination.

# Chapter 78

# AMARYLLIS: *HIPPEASTRUM HYBRIDUM* HORT.

## INTRODUCTION

Amaryllis is a popular, showy bulbous plant. The hybrids cultivated presently originated from crossing between *Hippeastrum reginae* and *H. vittatum.* These species are also grown in gardens for ornamental purposes. Amaryllis is planted in flower beds, borders, pots, window boxes, and used in landscaping and as cut flowers. Amaryllis grows up to 1 m in height and has strap-shaped leaves. The inflorescence is terminal and bears fragrant, trumpet-shaped flowers. Flowers form in a cluster of two to four and vary in color, from red, rose, white, and crimson, to orange and with red or white stripes. Fruit is globose and contains a few small green seeds. It is commonly propagated through the bulbs. Seeds are used for evolving suitable varieties through hybridization and for conservation of genetic diversity through seed storage.

## SEED STORAGE

### Seed Germination

Amaryllis seeds germinate better at alternate temperatures of 20/30°C for 16/8 h, respectively. Pindel  indicated that a temperature of 20°C was optimum for germination and that a higher temperature affected seedling emergence. Amico and colleagues reported that storing seeds at 4°C for 30 or 60 days increased germination speed and seedling emergence but decreased the total germination. The percentage of germination is higher in large seeds than in small seeds. Seed germination improves with seed stratification. Germination takes 28 days to complete.

## Storage Conditions

Amaryllis seeds are fairly good storers under ambient conditions and show orthodox storage behavior. High seed moisture and high storage temperature reduce the storage life of seeds. Carpenter and Ostmark reported that seed viability is unaffected by the dehydration of seeds. Seed storage at lower temperatures of 2 to 6°C maintains their viability for eight months. Seeds are tolerant to freezing, but viability decreases gradually as freezing temperatures decrease from 0 to –80°C. Seeds with low moisture content give a higher percentage of germination. Seeds having 11 percent moisture are viable for 12 months when stored at 5 or 15°C. Viability is affected by a higher storage temperature of 25°C or 35°C. Seed via bility is lost completely within three months of storage at higher temperatures (25 to 35°C) and higher relative humidity (75 to 95 percent).

# Chapter 79

# MARIGOLD (*TAGETES* SPP.)

## INTRODUCTION

*Figure.1. Marigold Plant in Bloom*

Marigold is a popular garden flower that blooms for a long period (seeFigure.1). It is also used as a cut flower and for decoration purposes due to its excellent keeping quality. Two types of marigolds, namely, African *(Tagetes erecta)* and French *(T. patula),* are commonly grown in gardens. The latter is dwarf in nature and bears profuse small flowers. This type is used in hanging baskets, window boxes, and rockeries. Its flowers are single or double and vary in shape, size, and color, such as yellow, orange, or red. It is a hardy plant that grows well in a wide range of climatic and soil conditions, and is commonly propagated from seeds. Occasionally, plants are also raised from cuttings. Furthermore, its seeds are mainly used in breeding programs for evolving new varieties and in the conservation of genetic diversity in the gene bank.

## SEED STORAGE

### Seed Germination

Fresh seeds exhibit dormancy and require about three months of afterripening before germination. Marigold seeds germinate better at a constant temperature of 20°C or at alternate temperatures of 20/30°C for 16/8 h, respectively, in the presence of light. Seed germination is rapid and good after soaking seeds in potassium nitrate (0.75 percent).

### Storage Conditions

Marigold seeds exhibit orthodox storage behavior and store fairly well under ambient conditions. Seed drying is beneficial in maintaining high viability during storage. The lowering of seed moisture and seed storage at low temperatures preserve viability for a longer period. Nikolova-Khristeva reported that seed moisture greater than 9 percent was injurious to seeds during storage, causing them to lose viability rapidly. Seeds with 8 percent moisture showed 38 to 50 percent germination during the first year of storage, and 30 to 39 percent during the second year. Storing seeds with 5 to 6 percent moisture in moisture-proof containers at lower temperature is beneficial for longer seed storage. Seeds of cultivars Queen Sophia and Butter Scotch are stored in polyethylene (0.175 mm) and laminated aluminum foil pouches at 5 and –20°C. Seed viability is lost within a year under ambient conditions, whereas it is preserved for three years at 5 and –20°C storage (Doijode, unpublished data). However, Ramakrishnan, Khan, and Khan reported complete loss of viability by 21 months under ambient conditions.

# Chapter 80

# DELPHINIUM (*DELPHINIUM* SPP.)

## INTRODUCTION

*Delphinium ajacis* and *D. consolida* are two commonly grown *Delphinium* species that are planted in gardens and also used as cut flowers. They are commonly planted in borders or beds. Flowers are blue, purple, pink, or white and arranged in spikes. Delphinium is propagated through seeds. Seeds are also used in evolving new cultivars through hybridization and also in the conservation of genetic diversity in the gene bank.

## SEED STORAGE

Seeds exhibit orthodox storage behavior and maintain high viability at low storage temperature and low relative humidity. Seed germination in delphinium is better at a constant temperature between 15 to 20°C or at alternate temperatures of 10/20°C or 15/25°C for 16/8 h. Delphinium seeds remain viable for a fairly long period under ambient conditions. Carpenter and Boucher  reported that delphinium seeds could tolerate storage at low, nonfreezing, or subzero temperatures. The highest seed germination occurred following storage at 5°C and 30 to 50 percent RH. Air-dried seeds stored at 21.1°C satisfactorily preserve viability, but germination decreases after 38 months of storage. Seeds gain moisture through frequent opening and closing of containers, thereby affecting seed viability. Seed viability preserves well for five years at 5°C and for ten years or more at –4°C.

# Chapter 81

# PETUNIA: *PETUNIA HYBRIDA* HORT.

## INTRODUCTION

Petunia is a popular annual flower grown in gardens. Plants flower profusely and continue to bloom for a long period. The present-day cultivars are derived from *Petunia integrifolia* and *P. nyctaginiflora.* Petunia is excellent for flower beds, hanging baskets, pots, and window boxes. Flowers are showy, solitary, axillary, and funnel shaped; they are white, yellow, bluish, pink, and red in color and fragrant. Fruit is a capsule containing numerous small seeds. Seeds are commonly used for propagation. They are also used in evolving new cultivars through hybridization and in the conservation of genetic diversity in the gene bank.

## SEED STORAGE

### Seed Germination

Petunia seeds germinate better at alternate temperatures of 20/30°C for 16/8 h in the presence of florescent light. Corbineau and Come  reported that petunia seeds germinate at between 15 to 35°C and the optimum temperature is 30°C. Jassey and colleagues  studied germination behavior in 25 lines of *Petunia* species and hybrids up to 50 weeks after harvest. Though seeds of a few lines germinated without any dormancy throughout storage, seeds of a few species did not germinate at 25 or 30°C unless treated with gibberellic acids or exposed to light. Seeds of *P. parodii* remained dormant even after 50 weeks of storage, as they contain inhibitors in the testae and germinate readily on isolation of embryos and treatment with gibberellic acid.

## Storage Conditions

Petunia seeds survive for a short period under ambient conditions, and they exhibit orthodox storage behavior. High seed viability is maintained under dry, cool conditions. Seeds stored in open boxes or paper packets lose viability considerably within four months and completely by nine months of storage under ambient conditions (3 to 30°C and 32 to 72 percent RH). Seed viability is preserved for three to four years in sealed plastic bags and also by additions of silica gel. Seeds stored in sealed containers under refrigerated conditions (5 to 7°C) retain viability for five years.

# Chapter 82

# GERBERA: *GERBERA JAMESONII ADLAM*

## INTRODUCTION

Gerbera is largely grown for its cut flowers but is also planted in rock gardens, flower beds, borders, and pots. It is a dwarf perennial herbaceous plant that grows up to 45 cm in height. Leaves form a cluster at the base. Flowers are single or double with perfect disc and ray flowers that can be yellow, salmon, pink, red, white, or orange (see Figure.2). Flowers have a vase life of 10 to 15 days. Gerbera prefers cool conditions for optimum growth. Seeds and division of clumps are used in propagation.

## SEED STORAGE

### Seed Germination

Gerbera seeds germinate at a constant temperature between 15 and 25°. However, Ellis, Hong, and Roberts reported that a constant temperature of 20°C or alternate temperatures of 20/30°C for 16/8 h are ideal for rapid and higher seed germination.

A longer germination period results from alternating temperatures. Seed germination is affected by reduction of moisture below 3.5 percent.

## Storage Conditions

Gerbera seeds are short-lived under ambient conditions, and they show orthodox storage behavior. Seeds preserve high viability and vigor for a longer period under dry, cool conditions. Carpenter, Ostmark, and Cornell reported that moisture reduction from 7.1 to 3.5 percent had no effect on germination. Seeds lose viability after nine months of storage under ambient conditions but preserve well for 30 months at –5 to 5°C. Carpenter, Ostmark, and Cornell observed similar germination percentage after storage at temperatures between 5 and –5°C, whereas germination was reduced when seeds were stored at –10°C or below. Seed viability can be preserved for 12 months at storage conditions of 5°C and 32 percent RH. Seeds stored at moderate relative humidity (52 percent) showed low viability, and none of them germinated when stored at high relative humidity (75 percent) and high temperature (25°C) for 12 months.

*Figure 1. Gerberas in the Flowering Stage*

# Chapter 83

# PERIWINKLE: *CATHARANTHUS ROSEUS* (L.) DON

## INTRODUCTION

Periwinkle is grown in pots, rockeries, window boxes, and as a ground cover. Plants are bushy and dwarf in habit, growing up to 60 cm in height. Leaves are oval and shiny. Plants flower profusely for a long time. Flowers are round or flat and white or pink in color. Seeds and cuttings are used for propagation. Seeds are also used in crop improvement through hybridization and also in the conservation of genetic diversity in the gene bank.

## SEED STORAGE

Periwinkle seeds are small, 2 to 3 mm in length and 1 to 2 mm in width. One thousand seeds weigh about 1 to 2 g. Fresh seeds readily germinate and exhibit 95 to 96 percent germination. Seed viability becomes zero after two years of storage under improper storage conditions. However, Mitrev reported that seed viability decreased gradually under ambient conditions. The germination percentage was 95, 53, 28, and 0 after one, two, three, and four years of storage, respectively, under ambient conditions. Seeds stored in desiccators over calcium chloride showed 49 percent germination after four years of storage. Periwinkle seeds retain viability for a fairly long period under ambient conditions and exhibit orthodox storage behavior. For longer seed storage, low-moisture seeds are packed in thick polyethylene bags or laminated aluminum foil pouches and stored at low temperatures.

# Chapter 84

# RHODODENDRON (*RHODODENDRON* SPP.)

## INTRODUCTION

Rhododendron is an evergreen or deciduous small perennial shrub grown in temperate zones for its beautiful flowers. Several *Rhododendron* species are grown in gardens for ornamental purposes, such as *R. indicum, R. japonicum, R. kotschyi, R. luteum, R. macgregoriae, R. obtusum, R. occidentale, R. simsii, R. sinense, R. vaseyi,* and *R. viscosum.* Flowers are borne terminally in clusters; they are large, single or double, and vary in color, such as pink, red, scarlet or carmine, and white. Seeds, cuttings, and grafts are used in propagation.

## SEED STORAGE

### Seed Germination

Rhododendron seeds exhibit dormancy. Fresh seeds of *R. kotschyi* showed 6 to 12 percent germination, reaching a maximum of 30 to 43 percent in 8 to 11 months of storage. The optimum temperature for germination is 20°C. Lee, Song, and Hong reported that seed washing with water adversely affected germination, whereas germination improved with seed soaking in thiourea (1 percent) and indolebutyric acid (100 ppm).

## Storage Conditions

Rhododendron seed viability is preserved for a short period under ambient conditions, particularly a few weeks at room temperatures. Seeds show orthodox storage behavior. Furthermore, deterioration decreases at lower storage temperatures and relative humidity. Germination of seeds of *R. macgregoriae* was reduced by 10 percent in two weeks when seeds were stored at 27°C and in 85 weeks at 4°C and less than 5 percent RH. Short seed life is not due to recalcitrance. In addition, such seeds dry well for longer storage at lower temperatures. Fillmore recommends storing rhododendron seeds in the refrigerator in sealed containers. Seeds of *R. kotschyi* preserved for four to five years show low germination percentage.

# Chapter 85

# ANTHURIUM (*ANTHURIUM* SPP.)

## INTRODUCTION

Anthurium is grown for its attractive foliage and beautiful, showy flowers. The *Anthurium* species most commonly grown for their foliage are *A. clarinervium, A. crystallinum, A. digitatum, A. magnificum, A. veitchii,* and *A. warocqueanum,* and the species commonly grown for their flowers are *A. andraeanum* and *A. scherzerianum.* Anthurium grows well in shade and in warm, humid conditions. It is also grown in pots for indoor gardening. Plants are dwarf and compact, with large, dark green, velvety, heart-shaped or ovate leaves. The inflorescence consists of a spathe, and the spadix is showy and covered with bisexual flowers. Fruits are striking and brilliantly colored. Seeds, suckers, and stem cuttings are commonly used in propagation. Seeds are also used in hybridization programs for crop improvement.

## SEED STORAGE

### Seed Germination

Anthurium seeds show low viability and germinate poorly under in vivo conditions. Embryos are small and poorly developed, causing delayed and poor germination. The excised embryos germinate better under in vitro conditions.

### Storage Conditions

Anthurium seeds exhibit recalcitrant storage behavior. They survive for an exceptionally short period under ambient conditions. Seeds lose viability rapidly on desiccation, and it is difficult to store them longer, even under moist conditions; furthermore, high seed moisture is favorable for the growth of pathogens. However, seed viability is preserved for a shorter period when seeds are stored in berries. Bachthaler reported that berries disinfected with benomyl, zineb, captan, and thiram maintained viability (60 percent) for 16 weeks at 10°C, whereas berries stored at 2°C suffered cold injury, and those stored at 5°C were affected by fungal infections and cold injury. Seed storage in a low-pressure cabinet (700 and 400 hectopascals [hPa]) preserved viability for 20 weeks at 11°C and gave 70 and 50 percent germination, respectively.

# Chapter 86

# CALADIUM: *CALADIUM BICOLOR* (AIT.) VENT.

## INTRODUCTION

Caladiums are predominantly grown in pots or in garden beds for their beautiful foliage. Leaves are broad and of various colors, such as red, rose, salmon, white, green, pink, and purple, in a speckled pattern. Seeds and bulbs are used in propagation. Seeds are also used in hybridization programs for evolving new genotypes and in the conservation of genetic diversity.

## SEED STORAGE

Caladium seeds germinate immediately after harvest, and they require continuous light for germination. Seed germination decreases from 95 to 87 percent when moisture content falls below 14 percent. These seeds appear to be recalcitrant in storage behavior and maintain high viability under cooler conditions. Caladium seeds retain high viability for seven days under a wide range of low temperatures (10 to –80°C) without loss of viability. Further, seeds are successfully preserved for six months at 15°C and 22 percent RH, without a change in initial viability.

However, seed viability will be significantly reduced at a higher temperature (25°C) and higher relative humidity (95 percent).

# Chapter 87

# GERANIUM (*PELARGONIUM* SPP.)

## INTRODUCTION

Geranium is grown for its beautiful foliage and flowers. The commonly grown geraniums are *Pelargonium peltatum, P. radens,* and *P. zonale.* Leaves are dark green and round. Flowers are round, single or double, and come in various colors, such as pink, scarlet, crimson, red, and purple.

Flowers bloom over a long period. Geranium is suitable for planting in pots, beds, or window boxes. Seeds and terminal cuttings are used in propagation.

## SEED STORAGE

Geranium seeds germinate better at alternate temperatures of 20/30°C for 16/8 h or at a constant temperature of 20°C, and germination takes 28 days to complete. Geranium seeds store fairly well under ambient conditions. They show orthodox storage behavior and deteriorate rapidly at higher storage temperatures and relative humidity. Seeds remain viable for a long period under cool, dry conditions. They can be preserved for five years when stored at 20°C, showing 100 percent germination.

# Chapter 88

# CAMELLIA (*CAMELLIA* SPP.)

## INTRODUCTION

Camellias are popular ornamental shrubs. Many of the existing varieties are hybrids of *Camellia japonica* and *C. saluenensis.* The plant is an evergreen, and the flowers are attractive, bisexual, single, axillary or terminal, funnel shaped, and white or pink. Camellia is commonly propagated by stem cuttings and grafting. Seeds are used for evolving new genotypes through hybridization.

## SEED STORAGE

Camellia seeds appear to be orthodox in storage behavior and retain high viability under cool, dry conditions. High seed viability can be preserved at low temperatures and low relative humidity. Seeds stored at 4.4°C maintain viability for one year. Freshly collected seeds should be kept for 18 days at room temperatures, then covered with plastic film and stored at 0 to 2°C to maintain high germination percentage for more than two years.

# REFERENCES


Ali, N. 1959. Studies on seed germination in rootstock species of *Citrus*. *Punjab Fruit J.* 22:22-32.

Bajpai, P.N., Trivedi, R.K., and Prasad, A. 1963. Storage of citrus seeds. *Sci. Cult.* 29:45-46.

Fucik, J.E. 1974. Grapefruit seed viability and germination. *J. Rio Grande Valley Hort. Soc.* 28:140-142.

Kadam, A.S., Khedkar, D.H., Patel, V.K., and Anserwadekar, K.W. 1994. Studies on viability and germinability of Rangpur lime seeds during storage. *J. Mah. Agril. Universities* 19:130-131.

Koyamu, K. 1951. A preliminary note on the germination of papaya seeds. *Madras nAgril. J.* 38:348-349.

Maeda, J.A., Pereira, M.De.F.D.A., and Terra, M.M. 1985. Storage conditions and grape vine seed viability and dormancy. *Bragantia* 44:245-254.

Singh, S. and Soni, S.L. 1974. Effect of water and acid soaking periods on seed germination in guava. *Punjab Hort. J.* 14:122-124.

Sinha, M.M., Verma, J.P., and Koranga, D.S. 1973. Studies on the seed germination of guava (*Psidium guavajava* L.). I. Effect of scarification and plant growth regulator treatments. *Prog. Hort.* 5:37-40.

Spalding, D.H., Knight, R.J., Jr., and Reeder, W.F. 1976. Storage of avocado seeds. *Proc. Fl. St. Hort. Soc.* 89:257-258.

Hayat, M.A. 1963. Morphology of seed germination and seedling in *Annona squamosa. Bot. Gaz.* 124:360-362.

Storey, W.B. 1969. Macadamia. In *Handbook of North America nut trees,* Jaynes R.A. (Ed.). Geneva,NY: The W.F. Humphery Press, Inc., pp. 321-335.

Ellis, R.H. 1984. Revised table of seed storage characteristics. *Plant Genetic Resources. Newsletter* 58:16-33.

Farooqui, A.A., Nalwadi, U.G., and Sulladmath, U.V. 1971. Effect of growth regulators on the germination of sapota *(Achras sapota)* seeds. *Mysore J. Agric. Sci.* 5:341-343.

Dutt, B.K. and Thakurta, A.G. 1956. Viability of vegetable seeds in storage. *Bose Res. Inst. Calcutta Tran.* 19:27-36.

Nath, P. 1976. *Vegetables for the tropical region.* New Delhi: Indian Council of Agriculture Research, p. 109.

# INDEX

www.ingramcontent.com/pod-product-compliance
Ingram Content Group UK Ltd.
Pitfield, Milton Keynes, MK11 3LW, UK
UKHW021328190726
13845UKWH00004B/78